T0214109

Introduction to Mechanics of Particles and Systems

Costas J. Papachristou

Introduction to Mechanics of Particles and Systems

 Springer

Costas J. Papachristou
Department of Physical Sciences
Hellenic Naval Academy
Piraeus, Greece

ISBN 978-3-030-54273-3   ISBN 978-3-030-54271-9   (eBook)
https://doi.org/10.1007/978-3-030-54271-9

This Springer imprint is published by the registered company Springer Nature Switzerland AG
The registered company address is: Gewerbestrasse 11, 6330 Cham, Switzerland

Preface

Newtonian Mechanics is, traditionally, the first stage of "initiation" of a college student into Physics. It is perhaps the only truly autonomous subject area of Physics, in the sense that it can be taught as a self-contained entity without the need for support from other areas of physical science. This textbook is based on lecture notes (originally in Greek) used by this author in his two-semester course of introductory Mechanics, taught at the Hellenic Naval Academy (the Naval Academy of Greece).

It is evident that no serious approach to Mechanics (at least at the university level) is possible without the support of higher Mathematics. Indeed, the central law of Mechanics, Newton's Second Law, carries a rich mathematical structure being both a vector equation and a differential equation. An effort is thus made to familiarize the student from the outset with the use of some basic mathematical tools, such as vectors, differential operators, and differential equations. To this end, the first chapter contains the elements of vector analysis that will be needed in the sequel, while the Mathematical Supplement constitutes a brief introduction to the aforementioned concepts of differential calculus.

The main text may be subdivided into three parts. In the first part (Chaps. 2–5), we study the mechanics of a single particle (and, more generally, of a body that executes purely translational motion) while the second part (Chaps. 6–8) introduces to the mechanics of more complex structures such as systems of particles, rigid bodies, and ideal fluids. The third part consists of 60 fully solved problems. I urge the student to try to solve each problem on his/her own before looking at the accompanying solution. Some useful supplementary material may also be found in the Appendices.

I am indebted to Aristidis N. Magoulas for helping me with the figures. I also thank the Hellenic Naval Academy for publishing the original, Greek version of the book. And, of course, I express my gratitude and appreciation to my wife, Thalia, for her patience and support while this book was written!

Piraeus, Greece   Costas J. Papachristou
June 2020

Contents

Chapter 1
Vectors

Abstract Fundamental concepts of vector analysis are presented. The scalar and vector products of vectors are defined.

1.1 Basic Notions

Vectors are physical or mathematical quantities carrying two properties: *magnitude* and *direction*. Symbolically, a vector is usually represented by an arrow (Fig. 1.1).

The magnitude of $\vec{V}$ (proportional, by convention, to the length of the arrow) is denoted by $\left|\vec{V}\right|$ and, by definition, $\left|\vec{V}\right| \geq 0$. In particular, a vector of zero magnitude is called a *zero vector*, $\vec{V} = 0$, and its direction is indeterminate. By definition, the vector $-\vec{V}$ has the same magnitude as $\vec{V}$ but is oriented in the opposite direction (Fig. 1.2).

A *unit vector* (denoted $\hat{u}$) is a vector of unit magnitude: $\left|\hat{u}\right| = 1$. A vector $\vec{V}$ in the direction of the unit vector $\hat{u}$ is written

$$\vec{V} = \left|\vec{V}\right|\hat{u}$$

while a vector $\vec{W}$ in the *opposite* direction is written

$$\vec{W} = -\left|\vec{W}\right|\hat{u} \ .$$

Note that the unit vector $\hat{u}$ in the direction of a vector $\vec{V}$ can be expressed as the quotient

$$\hat{u} = \frac{\vec{V}}{\left|\vec{V}\right|} \tag{1.1}$$

C. J. Papachristou, *Introduction to Mechanics of Particles and Systems*,
https://doi.org/10.1007/978-3-030-54271-9_1

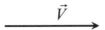

Fig. 1.1 Symbolic representation of a vector

Fig. 1.2 Two vectors having equal magnitudes and opposite directions

(By definition, the effect of multiplying or dividing a vector by a *positive* number is to multiply or divide, respectively, the magnitude of this vector by this number without altering the direction of the vector.) In general, a vector *parallel* to $\hat{u}$ is expressed as

$$\vec{V} = \pm \left| \vec{V} \right| \hat{u} \equiv V\,\hat{u} \tag{1.2}$$

The quantity V is called the *algebraic value* of $\vec{V}$ with respect to the unit vector $\hat{u}$. In particular, the sign of V indicates the orientation of $\vec{V}$ relative to $\hat{u}$.

Example: For the vectors in Fig. 1.3, we have:

$$\left| \vec{V} \right| = \left| \vec{W} \right| = 2\,, \quad \vec{V} = 2\,\hat{u}\,, \quad \vec{W} = -2\,\hat{u}\,, \quad V = 2\,, \quad W = -2$$

where the last two quantities represent algebraic values.

The *sum* $\vec{V} = \vec{A} + \vec{B}$ of two vectors can be represented graphically in two ways, as shown in Fig. 1.4 (note carefully the way the angle θ between the two vectors is specified in the left diagram).

The *difference* $\vec{W} = \vec{A} - \vec{B} \equiv \vec{A} + (-\vec{B})$ of two vectors is found graphically as seen in Fig. 1.5. In the figure on the right, note that the arrow of $\vec{A} - \vec{B}$ is directed toward the tip of $\vec{A}$. (Draw the vector $\vec{B} - \vec{A}$.)

It can be shown (see Exercise 1 at the end of the chapter) that

$$\left| \vec{A} \pm \vec{B} \right| = (A^2 + B^2 \pm 2AB\cos\theta)^{1/2} \tag{1.3}$$

Fig. 1.3 Two vectors of equal magnitudes may have opposite algebraic values

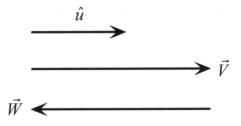

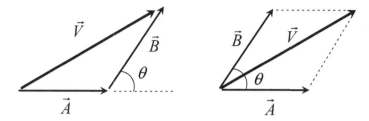

Fig. 1.4 Two equivalent graphic representations of the sum of two vectors: by placing the origin of either one at the tip of the other (left) or by forming a parallelogram with two vectors that have a common origin (right)

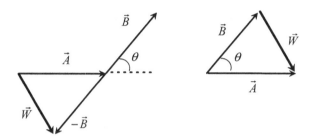

Fig. 1.5 Two equivalent graphic representations of the difference of two vectors: by adding the negative of the second vector to the first vector (left) or by joining the tips of two vectors with common origin, in the direction from the second vector to the first (right)

where $A = \left|\vec{A}\right|$, $B = \left|\vec{B}\right|$, and where θ is the angle between $\vec{A}$ and $\vec{B}$.

1.2 Rectangular Components of a Vector

An oriented straight line is called an *axis*. The orientation is specified by a unit vector $\hat{u}$, parallel to the line, the direction of which vector indicates the *positive* direction on the axis.

We consider the xy-plane defined by the mutually perpendicular x- and y-axes with unit vectors $\hat{u}_x$ and $\hat{u}_y$, respectively. Let $\vec{V}$ be a vector on this plane. It is often convenient to express $\vec{V}$ as a sum of two vectors parallel to the corresponding x- and y-axes, as seen in Fig. 1.6.

$$\vec{V} = \vec{V}_x + \vec{V}_y \text{ where } \vec{V}_x = V_x\,\hat{u}_x, \quad \vec{V}_y = V_y\,\hat{u}_y.$$

The quantities V_x and V_y, which are the *algebraic values* of the projections $\vec{V}_x$ and $\vec{V}_y$ of $\vec{V}$ onto the two axes, are the *rectangular components* of $\vec{V}$. These quantities may be positive or negative, depending on the orientations of $\vec{V}_x$ and $\vec{V}_y$ relative to $\hat{u}_x$ and $\hat{u}_y$, respectively. (In Fig. 1.6 both V_x and V_y are positive.) We write:

Fig. 1.6 Rectangular
components of a vector on
the *xy*-plane

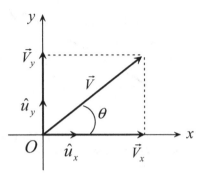

$$\vec{V} = V_x\,\hat{u}_x + V_y\,\hat{u}_y \equiv (V_x, V_y) \tag{1.4}$$

The *magnitude* of the vector $\vec{V}$ is

$$\left|\vec{V}\right| \equiv V = \sqrt{V_x^2 + V_y^2} \tag{1.5}$$

The angle θ in Fig. 1.6 is measured relative to the positive x semiaxis and it increases *counterclockwise*. Thus, starting from the x-axis we may form positive or negative angles by moving counterclockwise or clockwise, respectively. We note that

$$V_x = V\cos\theta\,, \qquad V_y = V\sin\theta\,, \qquad \tan\theta = \frac{V_y}{V_x} \tag{1.6}$$

In an analogous way one may define the rectangular components of a vector in 3-dimensional space. In this case we use a *right-handed* rectangular system of axes x, y, z, with corresponding unit vectors $\hat{u}_x, \hat{u}_y, \hat{u}_z$, as shown in Fig. 1.7. (If we interchanged, say, the names of the x- and y-axes, the ensuing rectangular system xyz would be *left-handed*, while the system yxz would now be right-handed. Can you find a practical way to determine whether a given system of axes xyz is right-handed or left-handed? Notice the *order* in which the names x, y and z are written!) The algebraic values V_x, V_y, V_z of the projections of $\vec{V}$ onto the three axes constitute the *rectangular components* of $\vec{V}$. The 3-dimensional generalizations of (1.4) and (1.5) are

$$\vec{V} = V_x\,\hat{u}_x + V_y\,\hat{u}_y + V_z\,\hat{u}_z \equiv (V_x, V_y, V_z)\,;$$
$$\left|\vec{V}\right| \equiv V = \sqrt{V_x^2 + V_y^2 + V_z^2} \tag{1.7}$$

In particular, if $\vec{V} = 0$ then $V_x = V_y = V_z = 0$ and $V = 0$.

Now, assume that $\vec{A} = A_x\hat{u}_x + A_y\hat{u}_y + A_z\hat{u}_z$ and $\vec{B} = B_x\hat{u}_x + B_y\hat{u}_y + B_z\hat{u}_z$. Then,

Fig. 1.7 A vector in
3-dimensional space

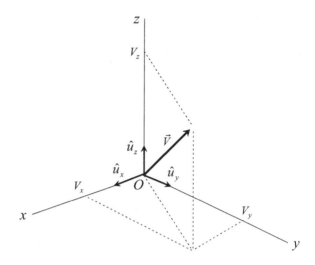

$$\vec{A} = \vec{B} \quad \Leftrightarrow \quad A_x = B_x, \quad A_y = B_y, \quad A_z = B_z \tag{1.8}$$

Moreover,

$$\begin{aligned} \vec{A} \pm \vec{B} &= (A_x \pm B_x)\hat{u}_x + (A_y \pm B_y)\hat{u}_y + (A_z \pm B_z)\hat{u}_z \\ &\equiv (A_x \pm B_x, \ A_y \pm B_y, \ A_z \pm B_z) \end{aligned} \tag{1.9}$$

In general, the components of a sum of vectors are the sums of the respective components of the vectors. Thus, let

$$\vec{V} = \vec{V}_1 + \vec{V}_2 + \cdots \equiv \sum_i \vec{V}_i \equiv (V_x, \ V_y, \ V_z) \text{ where } \vec{V}_i \equiv (V_{ix}, \ V_{iy}, \ V_{iz}).$$

Then,

$$V_x = \sum_i V_{ix}, \quad V_y = \sum_i V_{iy}, \quad V_z = \sum_i V_{iz} \tag{1.10}$$

Example: For the vectors $\vec{A} \equiv (1, -1, 0)$, $\vec{B} \equiv (2, 1, -1)$, we have:

$$\vec{A} + \vec{B} \equiv (3, 0, -1), \quad \vec{A} - \vec{B} \equiv (-1, -2, 1)$$

and

$$\left|\vec{A} + \vec{B}\right| = \sqrt{3^2 + 0 + (-1)^2} = \sqrt{10}, \quad \left|\vec{A} - \vec{B}\right| = \sqrt{(-1)^2 + (-2)^2 + 1^2} = \sqrt{6}.$$

1.3   Position Vectors

A *position vector* is a vector used to determine the position of a point in space, relative to a fixed reference point O which, typically, is chosen to be the origin of our coordinate system. For points on a plane, we use a system of two axes x, y (Fig. 1.8).

The vector $\vec{r} = \overrightarrow{OP}$ determines the position of point P relative to O. The components (x, y) of $\vec{r}$ are the *Cartesian coordinates* of P. We write:

$$\vec{r} = x\hat{u}_x + y\hat{u}_y \equiv (x, y) \tag{1.11}$$

Alternatively, we can determine the position of P by using *polar coordinates* (r, θ), where $r = |\vec{r}|$ and where $0 \leq \theta < 2\pi$ or $-\pi < \theta \leq \pi$ (by convention, the angle θ increases *counterclockwise*). We notice that

$$x = r \cos\theta \,, \quad y = r \sin\theta \tag{1.12}$$

or, conversely,

$$r = \sqrt{x^2 + y^2} \,, \quad \tan\theta = \frac{y}{x} \tag{1.13}$$

For points in space we use a rectangular system of axes x, y, z (Fig. 1.9). We write:

$$\vec{r} = x\hat{u}_x + y\hat{u}_y + z\hat{u}_z \equiv (x, y, z) \,;$$
$$|\vec{r}| = r = \sqrt{x^2 + y^2 + z^2} \tag{1.14}$$

The three quantities (x, y, z) constitute the *Cartesian coordinates* of point P in space. Alternative systems of coordinates are *spherical* and *cylindrical* coordinates, which will not be used in this book (see [1, 2]).

Now, let P_1, P_2 be two points in space with position vectors $\vec{r}_1$, $\vec{r}_2$ and with coordinates $(x_1, y_1, z_1), (x_2, y_2, z_2)$, respectively (Fig. 1.10). We seek an expression for the *distance* P_1P_2 between these points.

Fig. 1.8 A position vector on the *xy*-plane

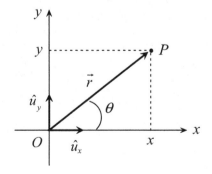

Fig. 1.9 A position vector in 3-dimensional space

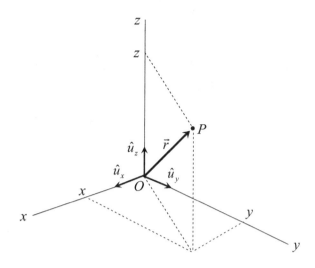

Fig. 1.10 The distance between two points in space is the magnitude of the difference of their position vectors

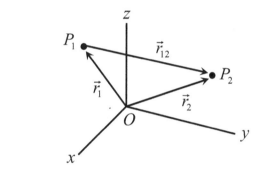

We notice that $P_1 P_2 = |\vec{r}_{12}| = |\vec{r}_2 - \vec{r}_1|$. But,

$$\vec{r}_1 = x_1 \hat{u}_x + y_1 \hat{u}_y + z_1 \hat{u}_z , \quad \vec{r}_2 = x_2 \hat{u}_x + y_2 \hat{u}_y + z_2 \hat{u}_z$$

so that

$$\vec{r}_{12} = \vec{r}_2 - \vec{r}_1 = (x_2 - x_1)\hat{u}_x + (y_2 - y_1)\hat{u}_y + (z_2 - z_1)\hat{u}_z .$$

Therefore,

$$P_1 P_2 = |\vec{r}_{12}| = \left[(x_2 - x_1)^2 + (y_2 - y_1)^2 + (z_2 - z_1)^2\right]^{1/2} \quad (1.15)$$

Example: For the points P_1, P_2, with respective coordinates $(x_1, y_1, z_1) \equiv (-2, 1, -3)$ and $(x_2, y_2, z_2) \equiv (0, -1, -2)$, we have: $P_1 P_2 = \sqrt{2^2 + (-2)^2 + 1^2} = 3$.

1.4 Scalar ("Dot") Product of Two Vectors

Consider two vectors $\vec{A}$ and $\vec{B}$, and let θ be the angle between them, where, by convention, $0 \leq \theta \leq \pi$ (Fig. 1.11). The *scalar product* (or "dot product") of $\vec{A}$ and $\vec{B}$ is a scalar quantity defined by the equation

$$\vec{A} \cdot \vec{B} = \left|\vec{A}\right|\left|\vec{B}\right| \cos\theta = AB\cos\theta \tag{1.16}$$

where $A = \left|\vec{A}\right|$, $B = \left|\vec{B}\right|$. It is easy to show that this product is *commutative*:

$$\vec{A} \cdot \vec{B} = \vec{B} \cdot \vec{A} \ .$$

In the case where $\vec{A} = \vec{B}$, we have $\theta = 0$, $\cos\theta = 1$ and

$$\vec{A} \cdot \vec{A} = \left|\vec{A}\right|^{2} = A^{2} \tag{1.17}$$

If $\vec{A}$ and $\vec{B}$ are *mutually perpendicular*, then $\theta = \pi/2$, $\cos\theta = 0$ and therefore

$$\vec{A} \cdot \vec{B} = 0 \quad \Leftrightarrow \quad \vec{A} \perp \vec{B} \tag{1.18}$$

As can be shown [1],

$$\vec{A} \cdot (\lambda\vec{B}) = (\lambda\vec{A}) \cdot \vec{B}, (\kappa\vec{A}) \cdot (\lambda\vec{B}) = \kappa\lambda(\vec{A} \cdot \vec{B})$$

(where κ, λ are scalars) and

$$\vec{A} \cdot (\vec{B} + \vec{C}) = \vec{A} \cdot \vec{B} + \vec{A} \cdot \vec{C} \ .$$

For the unit vectors we can show that

$$\hat{u}_x \cdot \hat{u}_x = \hat{u}_y \cdot \hat{u}_y = \hat{u}_z \cdot \hat{u}_z = 1 \ , \qquad \hat{u}_x \cdot \hat{u}_y = \hat{u}_x \cdot \hat{u}_z = \hat{u}_y \cdot \hat{u}_z = 0 \ .$$

Fig. 1.11 Two vectors forming an angle θ

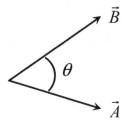

Thus, if $\vec{A} = A_x\hat{u}_x + A_y\hat{u}_y + A_z\hat{u}_z$ and $\vec{B} = B_x\hat{u}_x + B_y\hat{u}_y + B_z\hat{u}_z$, we find a useful relation for the dot product in terms of components:

$$\vec{A} \cdot \vec{B} = A_x B_x + A_y B_y + A_z B_z \qquad (1.19)$$

For two equal vectors we are thus led back to (1.17):

$$\vec{A} \cdot \vec{A} = A_x^2 + A_y^2 + A_z^2 = \left|\vec{A}\right|^2 \qquad (1.20)$$

1.5 Vector ("Cross") Product of Two Vectors

The *vector product* (or "cross product") $\vec{A} \times \vec{B}$ of $\vec{A}$ and $\vec{B}$ is a vector normal to the plane defined by $\vec{A}$ and $\vec{B}$ and in the direction of advance of a right-handed screw rotated from $\vec{A}$ to $\vec{B}$ (see Fig. 1.12). Alternatively, the direction of $\vec{A} \times \vec{B}$ may be determined by the "right-hand rule": Let the fingers of the right hand point in the direction of the first vector, $\vec{A}$. Rotate the fingers from $\vec{A}$ to $\vec{B}$ through the smaller of the two angles between these vectors. The extended thumb then points in the direction of $\vec{A} \times \vec{B}$. The magnitude of $\vec{A} \times \vec{B}$ is, by definition,

$$\left|\vec{A} \times \vec{B}\right| = \left|\vec{A}\right|\left|\vec{B}\right| \sin\theta = AB \sin\theta \qquad (1.21)$$

where $0 \leq \theta \leq \pi$. We notice that

$$\vec{A} \times \vec{B} = -\vec{B} \times \vec{A}$$

Fig. 1.12 The two possible orientations of the vector product of two vectors, depending on the order in which these vectors appear in the product

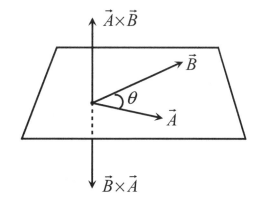

(hence, the cross product is *not* commutative) and that $\vec{A} \times \vec{A} = 0$. In general, if $\vec{A}$ and $\vec{B}$ are *parallel* vectors, then $\vec{A} \times \vec{B} = 0$, since, in this case, $\theta = 0$ or π, so that $\sin\theta = 0$.

As can be proven [1], $\vec{A} \times (\lambda \vec{B}) = (\lambda \vec{A}) \times \vec{B}$, $(\kappa \vec{A}) \times (\lambda \vec{B}) = \kappa\lambda(\vec{A} \times \vec{B})$

(where κ, λ are scalars) and

$$\vec{A} \times (\vec{B} + \vec{C}) = (\vec{A} \times \vec{B}) + (\vec{A} \times \vec{C}) \ .$$

For the unit vectors of a right-handed rectangular system we can show that

$$\hat{u}_x \times \hat{u}_x = \hat{u}_y \times \hat{u}_y = \hat{u}_z \times \hat{u}_z = 0, \quad \hat{u}_x \times \hat{u}_y = \hat{u}_z, \quad \hat{u}_y \times \hat{u}_z = \hat{u}_x, \quad \hat{u}_z \times \hat{u}_x = \hat{u}_y \ .$$

Thus, if $\vec{A} = A_x\hat{u}_x + A_y\hat{u}_y + A_z\hat{u}_z$, $\vec{B} = B_x\hat{u}_x + B_y\hat{u}_y + B_z\hat{u}_z$, we find that

$$\vec{A} \times \vec{B} = (A_y B_z - A_z B_y)\,\hat{u}_x + (A_z B_x - A_x B_z)\,\hat{u}_y + (A_x B_y - A_y B_x)\,\hat{u}_z \quad (1.22)$$

This can be written more compactly in the form of a determinant,

$$\vec{A} \times \vec{B} \ = \ \begin{vmatrix} \hat{u}_x & \hat{u}_y & \hat{u}_z \\ A_x & A_y & A_z \\ B_x & B_y & B_z \end{vmatrix} \tag{1.23}$$

to be developed with respect to the first row.

Exercises

1. Consider the vectors $\vec{A} \equiv (A_x, A_y, A_z)$ and $\vec{B} \equiv (B_x, B_y, B_z)$, and let θ be the angle between them. By using the properties of the scalar product, show the following:

   a. $\left| \vec{A} \pm \vec{B} \right| = \left(\left| \vec{A} \right|^2 + \left| \vec{B} \right|^2 \pm 2 \left| \vec{A} \right| \left| \vec{B} \right| \cos\theta \right)^{1/2}$

   b. $\cos\theta = \dfrac{A_x B_x + A_y B_y + A_z B_z}{\left(A_x^2 + A_y^2 + A_z^2\right)^{1/2} \left(B_x^2 + B_y^2 + B_z^2\right)^{1/2}}$

   c. If $\vec{A} \perp \vec{B}$, then $\left| \vec{A} + \vec{B} \right|^2 = \left| \vec{A} - \vec{B} \right|^2 = \left| \vec{A} \right|^2 + \left| \vec{B} \right|^2$ (*Pythagorean theorem*)

2. Let $\vec{A} \equiv (A_x, A_y, A_z)$, $\vec{B} \equiv (B_x, B_y, B_z)$, $\vec{C} \equiv (C_x, C_y, C_z)$.

   a. Show that

   $$\vec{A} \cdot \left(\vec{B} \times \vec{C} \right) = \begin{vmatrix} A_x & A_y & A_z \\ B_x & B_y & B_z \\ C_x & C_y & C_z \end{vmatrix}$$

b. By using the properties of determinants, show that

$$\vec{A} \cdot (\vec{B} \times \vec{C}) = \vec{B} \cdot (\vec{C} \times \vec{A}) = \vec{C} \cdot (\vec{A} \times \vec{B})$$
$$\vec{A} \cdot (\vec{A} \times \vec{B}) = 0$$

3. Find the value of α in order that the vectors $\vec{A} \equiv \left(\frac{1}{2}, \alpha, \frac{3}{2}\right)$ and $\vec{B} \equiv (-3, 3, -1)$ be mutually perpendicular.

4. Find the values of α and β in order that the vectors $\vec{A} \equiv (1, \alpha, 3)$ and $\vec{B} \equiv (-2, -4, \ \beta)$ be parallel to each other.

References

1. A.I. Borisenko, I.E. Tarapov, *Vector and Tensor Analysis with Applications* (Dover, 1979)
2. M.D. Greenberg, *Advanced Engineering Mathematics*, 2nd edn. (Prentice-Hall, 1998)

Chapter 2
Kinematics

Abstract Motion per se is studied, independently of the physical factors that cause or affect it. The vector concepts of velocity and acceleration of a particle are defined. Circular motion and relative motion are studied.

2.1 Rectilinear Motion

Kinematics is the branch of Mechanics that studies motion per se, regardless of the physical factors that cause or affect it. (The connection between cause and effect is the subject of *Dynamics*, to be discussed in the next chapter.)

The simplest type of motion is *rectilinear motion*, i.e., motion along a straight line. Such a line could be, e.g., the x-axis on which we have defined a positive orientation in the direction of the unit vector $\hat{u}_x$, as well as a point O (an *origin*) at which $x = 0$ (Fig. 2.1).

The position A of the moving object, at time t, is specified by the position vector

$$\overrightarrow{OA} = \vec{r} = x\hat{u}_x$$

where x and $\vec{r}$ are functions of t: $x = x(t)$, $\vec{r} = \vec{r}(t)$. We note that $x > 0$ or $x < 0$, depending on whether the object is on the right or on the left of O, respectively.

The *velocity* of the object at point A, at time t, is defined as the time derivative of the position vector; i.e., it is the vector

$$\vec{v} = \frac{d\vec{r}}{dt} = \frac{d}{dt}(x\hat{u}_x) = \frac{dx}{dt}\hat{u}_x$$

(where we have taken into account that $\hat{u}_x$ is constant). We write:

$$\vec{v} = v\hat{u}_x \quad \text{where} \quad v = \frac{dx}{dt} = \pm|\vec{v}| \tag{2.1}$$

© The Editor(s) (if applicable) and The Author(s), under exclusive license to Springer Nature Switzerland AG 2020
C. J. Papachristou, *Introduction to Mechanics of Particles and Systems*,
https://doi.org/10.1007/978-3-030-54271-9_2

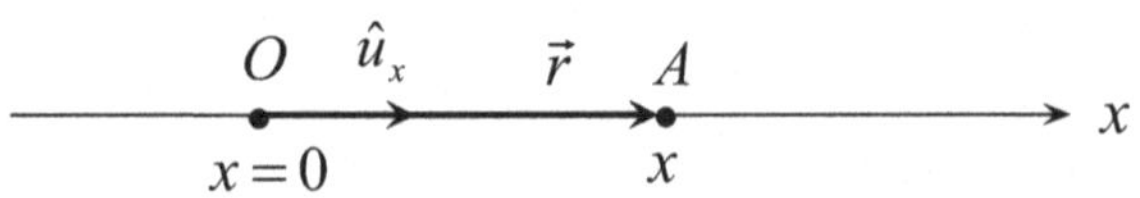

Fig. 2.1 Motion along the x-axis; the value of x determines the momentary position A of the moving object

In the above relation, v is the *algebraic value* of the velocity with respect to the unit vector $\hat{u}_x$. The magnitude of the velocity, equal to $|v|$, is called the *speed*. In general, v and $\vec{v}$ are functions of t. The *sign* of v indicates the instantaneous *direction* of motion: if $v > 0$, the object is moving in the positive direction of the x-axis (i.e., to the right, as seen in Fig. 2.1), while if $v < 0$, the object is moving in the negative direction (to the left). In S.I. units, v is expressed in $m/s = m.s^{-1}$.

Given the function $v = v(t)$, we can find the position $x(t)$ of the object at all times t by integrating (cf. Mathematical Supplement). From (2.1) we have:

$$v = \frac{dx}{dt} \quad \Rightarrow \quad dx = v\, dt \,.$$

We integrate the above differential equation, making the additional assumption (*initial condition*) that, at the moment $t = t_0$, the instantaneous position of the object is $x = x_0$:

$$\int_{x_0}^{x} dx = \int_{t_0}^{t} v\, dt \;\Rightarrow\; x - x_0 = \int_{t_0}^{t} v\, dt \;\Rightarrow$$

$$x = x_0 + \int_{t_0}^{t} v\, dt \tag{2.2}$$

The difference $x - x_0$ is called the *displacement* of the object from point x_0.

The *acceleration* of the object at time t is the derivative of the velocity vector:

$$\vec{a} = \frac{d\vec{v}}{dt} = \frac{d}{dt}(v\hat{u}_x) = \frac{dv}{dt}\hat{u}_x \,.$$

We write:

$$\vec{a} = a\hat{u}_x \quad \text{where} \quad a = \frac{dv}{dt} = \frac{d^2x}{dt^2} = \pm|\vec{a}| \tag{2.3}$$

The unit of acceleration in the S.I. system is $m/s^2 = m\ s^{-2}$.

The quantity a in (2.3) represents the algebraic value of the acceleration. For a given function $a = a(t)$, and by assuming that at the moment $t = t_0$ the moving object has a velocity $v = v_0$, we find the velocity $v(t)$ at all times t by integrating:

$$a = \frac{dv}{dt} \quad \Rightarrow \quad dv = a\,dt \quad \Rightarrow \quad \int_{v_0}^{v} dv = \int_{t_0}^{t} a\,dt \quad \Rightarrow$$

$$v = v_0 + \int_{t_0}^{t} a\,dt \tag{2.4}$$

If we know the acceleration as a function of x, $a = a(x)$, we can find the velocity as a function of position, as follows: By dividing the relations $dv = a\,dt$ and $dx = v\,dt$ in order to eliminate t, we get: $v\,dv = a\,dx$. We now integrate this relation, assuming that $v = v_0$ at the position $x = x_0$:

$$\int_{v_0}^{v} v\,dv = \int_{x_0}^{x} a\,dx \quad \Rightarrow \quad \frac{v^2}{2} - \frac{v_0^2}{2} = \int_{x_0}^{x} a\,dx \quad \Rightarrow$$

$$v^2 = v_0^2 + 2 \int_{x_0}^{x} a\,dx \tag{2.5}$$

Two observations are in order regarding Eq. (2.5):

1. In order for (2.5) to make sense physically, the right-hand side must be *nonnegative*. This may place restrictions on the admissible values of x, which means that the object may not be allowed to move on the entire x-axis.
2. To find v itself (rather than just its absolute value) from relation (2.5), we must take the square root of the right-hand side (the latter assumed nonnegative). This process will yield two values for v, with opposite signs. One of these values must be excluded, however, since its sign will not be consistent with that of v_0.

Note also that, in relations (2.2) and (2.4), the x and v are functions of the variable t that appears *on the upper limits* of the corresponding integrals. Similarly, in relation (2.5) the quantity v^2 is a function of the variable x that appears on the upper limit of the integral. In general, an integral with variable upper limit is a function of that limit [1, 2].

2.2 Special Types of Rectilinear Motion

We now apply the general results of the previous section to two familiar cases of rectilinear motion.

1. *Uniform rectilinear motion: $v = $ constant, $a = 0$*

This is the motion with *constant velocity* (in magnitude and direction). Relation (2.2) yields (by putting $t_0 = 0$):

$$x = x_0 + \int_{0}^{t} v\,dt = x_0 + v \int_{0}^{t} dt \quad \Rightarrow$$

$$\boxed{x = x_0 + vt} \tag{2.6}$$

2. *Uniformly accelerated rectilinear motion:* $a = constant \neq 0$

By relations (2.4) and (2.2) we find (putting $t_0 = 0$):

$$v = v_0 + \int_0^t a \, dt = v_0 + a \int_0^t dt \;\Rightarrow$$

$$\boxed{v = v_0 + at} \tag{2.7}$$

$$x = x_0 + \int_0^t v \, dt = x_0 + \int_0^t (v_0 + at) \, dt = x_0 + v_0 \int_0^t dt + a \int_0^t t \, dt \;\Rightarrow$$

$$\boxed{x = x_0 + v_0 t + \frac{1}{2} at^2} \tag{2.8}$$

Furthermore, relation (2.5) gives:

$$v^2 = v_0^2 + 2 \int_{x_0}^x a \, dx = v_0^2 + 2a \int_{x_0}^x dx \;\Rightarrow$$

$$\boxed{v^2 = v_0^2 + 2a(x - x_0)} \tag{2.9}$$

Note that the same result is found by eliminating t between (2.7) and (2.8). Note also that, since the right-hand side of (2.9) must be nonnegative, the acceptable values of x may be restricted; this, in turn, will place a restriction on the possible positions of the moving object.

Example: Projectile motion

At time $t_0 = 0$, a bullet is fired straight upward from a point O (at which $x_0 = 0$) of the vertical x-axis, with initial velocity $\vec{v}_0 = v_0 \, \hat{u}_x$ ($v_0 > 0$) (Fig. 2.2). We assume that the bullet is subject only to the force of gravity (we ignore air resistance). Thus, the acceleration of the projectile equals the acceleration of gravity, which is always directed *downward, regardless of the direction of motion* (upward or downward) of the projectile. In vector form,

$$\vec{a} = -g \, \hat{u}_x \equiv a \, \hat{u}_x \;\Rightarrow\; a = -g$$

where g is approximately equal to 9.8 m/s^2. We notice that a is constant, and therefore the motion is *uniformly accelerated*. The equations of motion are:

$$v = v_0 + at = v_0 - gt \; ,$$

$$x = x_0 + v_0 t + \frac{1}{2} at^2 = v_0 t - \frac{1}{2} gt^2 \; ,$$

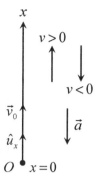

Fig. 2.2 Projectile motion along the vertical x-axis

$$v^2 = v_0^2 + 2a\,(x - x_0) = v_0^2 - 2gx \ .$$

The projectile will reach a maximum height $x = h$, where it will stop momentarily at time $t = t_h$; it will then start moving downward, toward the point of ejection O. To find h and t_h, we use the first two equations of motion:

$$v_h = v_0 - gt_h = 0 \ \Rightarrow \ t_h = \frac{v_0}{g} \ ;$$

$$h = v_0\,t_h - \frac{1}{2}gt_h^2 = \frac{v_0^2}{2g} \ .$$

The maximum height h may also be determined by noting that, in order for the right-hand side of the third equation of motion to be nonnegative, the value of x must not exceed $v_0^2/2g$. Note also that $v > 0$ for $0 < t < t_h$, while $v < 0$ for $t > t_h$. What is the physical meaning of this? (Notice that we chose the positive direction of the x-axis upward.)

Exercise: Find the moment at which the bullet returns to O, as well as the velocity of return. What do you observe? Also, show that, in a free fall from height h, a body acquires a speed

$$v = \sqrt{2gh} \ .$$

2.3 Curvilinear Motion in Space

We now consider motion along an arbitrary curve on the plane or in space. The instantaneous position of the moving object is determined by the position vector $\vec{r}$ with respect to the origin O of our coordinate system (Fig. 2.3). This vector, as well as the corresponding coordinates (x, y, z), are functions of time t.

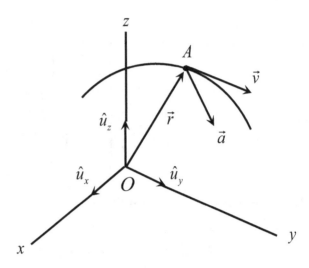

According to (1.14), we may write:

$$\vec{r} = \vec{r}(t) = x\hat{u}_x + y\hat{u}_y + z\hat{u}_z \quad \text{where} \quad x = x(t) \,, \quad y = y(t) \,, \quad z = z(t) \,.$$

The *velocity* of the moving object at a point A of the trajectory, at time t, is the time derivative of the position vector:

$$\vec{v} = \frac{d\vec{r}}{dt} = \frac{d}{dt}(x\hat{u}_x + y\hat{u}_y + z\hat{u}_z) = \frac{dx}{dt}\hat{u}_x + \frac{dy}{dt}\hat{u}_y + \frac{dz}{dt}\hat{u}_z \qquad (2.10)$$

We write:

$$\boxed{\begin{array}{c} \vec{v} = v_x\hat{u}_x + v_y\hat{u}_y + v_z\hat{u}_z \\ v_x = \frac{dx}{dt}, \quad v_y = \frac{dy}{dt}, \quad v_z = \frac{dz}{dt} \end{array}} \qquad (2.10')$$

As will be shown analytically in Sect. 2.6, the velocity vector $\vec{v}$ is *tangent* to the trajectory at all points and its direction is that of the direction of motion. The magnitude of the velocity, called *speed*, is

$$|\vec{v}| = \sqrt{v_x^2 + v_y^2 + v_z^2} \qquad (2.11)$$

The *acceleration* of the object at a point A, at time t, is the time derivative of the velocity:

$$\vec{a} = \frac{d\vec{v}}{dt} = \frac{dv_x}{dt}\hat{u}_x + \frac{dv_y}{dt}\hat{u}_y + \frac{dv_z}{dt}\hat{u}_z \qquad (2.12)$$

We write:

$$\boxed{\begin{array}{c} \vec{a} = a_x \hat{u}_x + a_y \hat{u}_y + a_z \hat{u}_z \\ a_x = \frac{dv_x}{dt}, \quad a_y = \frac{dv_y}{dt}, \quad a_z = \frac{dv_z}{dt} \end{array}} \tag{2.12'}$$

As will be shown in Sect. 2.6, the direction of the acceleration vector $\vec{a}$ is toward the *concave* ("inner") side of the trajectory. The magnitude of the acceleration is

$$|\vec{a}| = \sqrt{a_x^2 + a_y^2 + a_z^2} \tag{2.13}$$

Example: Assume that the coordinates of a moving particle are given as functions of time by the equations $\{x = A \cos \omega t, y = A \sin \omega t, z = \lambda t\}$, where A, ω, λ are positive constants. (What can you say about the trajectory of the particle?) By using (2.10′) and (2.12′) we find the components of velocity and acceleration, respectively, of the particle:

$$\left(v_x, v_y, v_z\right) \equiv \left(-\omega A \sin \omega t, \omega A \cos \omega t, \lambda\right),$$
$$\left(a_x, a_y, a_z\right) \equiv \left(-\omega^2 A \cos \omega t, -\omega^2 A \sin \omega t, 0\right).$$

The magnitudes v and a of the corresponding vectors are given by (2.11) and (2.13):

$$v = \sqrt{\omega^2 A^2 + \lambda^2}, \quad a = \omega^2 A.$$

Note that the speed v of the particle is constant in time, as well as that the vectors of velocity and acceleration are mutually perpendicular [show this by using relations (1.18) and (1.19)]. As will be seen below, these two facts are closely related.

2.4 Change of Speed

Generally speaking, a motion is *accelerated* if $\vec{a} \neq 0$, so that the vector $\vec{v}$ of the velocity changes with time. The term "*accelerated*", however, is often used with a different meaning in Kinematics, a fact that, if not properly pointed out, may lead to confusion. Thus, a (generally curvilinear) motion is said to be "*accelerated*" or "*retarded*" during a time interval if the *speed* $v = |\vec{v}|$ increases or decreases, respectively, in that interval. If the speed is constant, the motion is called *uniform*.

The kind of motion depends on the angle θ between the vector $\vec{v}$ of the velocity and the vector $\vec{a}$ of the acceleration, where, by convention, $0 \leq \theta \leq \pi$. In general,

$$\vec{v} \cdot \vec{a} = |\vec{v}||\vec{a}| \cos \theta = v \, a \, \cos \theta \tag{2.14}$$

where $a = |\vec{a}|$. On the other hand,

$$2(\vec{v} \cdot \vec{a}) = 2\left(\vec{v} \cdot \frac{d\vec{v}}{dt}\right) = \vec{v} \cdot \frac{d\vec{v}}{dt} + \frac{d\vec{v}}{dt} \cdot \vec{v} = \frac{d}{dt}(\vec{v} \cdot \vec{v}) = \frac{d(v^2)}{dt} = \frac{d(v^2)}{dv}\frac{dv}{dt} = 2v\frac{dv}{dt} \quad \Rightarrow$$

$$\vec{v} \cdot \vec{a} = \vec{v} \cdot \frac{d\vec{v}}{dt} = v\frac{dv}{dt} \tag{2.15}$$

where we have used (1.17), and where it should be noted carefully that

$$v\frac{dv}{dt} \equiv |\vec{v}|\frac{d|\vec{v}|}{dt} \neq |\vec{v}|\left|\frac{d\vec{v}}{dt}\right| \quad !$$

By comparing (2.14) and (2.15), we find that

$$\frac{dv}{dt} = a \cos\theta \tag{2.16}$$

Given that $a > 0$, we note the following:

a. If $0 \leq \theta < \frac{\pi}{2}$ then $\frac{dv}{dt} > 0$; v *increases* and the motion is *accelerated*.
b. If $\frac{\pi}{2} < \theta \leq \pi$ then $\frac{dv}{dt} < 0$; v *decreases* and the motion is *retarded*.
c. If $\theta = \frac{\pi}{2}$ (that is, if $\vec{a} \perp \vec{v}$) then $\frac{dv}{dt} = 0$; v is *constant* and the motion is *uniform*.

Of special importance is the following conclusion:

> If the acceleration is perpendicular to the velocity, the speed of the moving object is constant in time, even though the direction of the velocity is changing. Thus, the motion is uniform.

In the case of *rectilinear* motion, the angle θ between $\vec{v}$ and $\vec{a}$ can only assume two values. If $\vec{v}$ and $\vec{a}$ are in the *same* direction, then $\theta = 0$ and the motion is *accelerated*, while if $\vec{v}$ and $\vec{a}$ are in *opposite* directions, then $\theta = \pi$ and the motion is *retarded*. Now, $\vec{v}$ and $\vec{a}$ are in the same direction or in opposite directions if $\vec{v} \cdot \vec{a} > 0$ or $\vec{v} \cdot \vec{a} < 0$, respectively. Given that $\vec{v} = v\,\hat{u}_x$ and $\vec{a} = a\,\hat{u}_x$ [see Eqs. (2.1) and (2.3)], where here v and a are *algebraic values* ($v = \pm|\vec{v}|$, $a = \pm|\vec{a}|$), we have: $\vec{v} \cdot \vec{a} = v\,a$. We thus conclude that

> a rectilinear motion is accelerated or retarded, depending on whether the product of the algebraic values of the velocity and the acceleration is positive ($va > 0$) or negative ($va < 0$), respectively.

2.5 Motion with Constant Acceleration

We now consider the case where the acceleration $\vec{a}$ of the moving object is constant in magnitude and direction. We assume that at time $t = 0$ the instantaneous position vector of the object, relative to our coordinate system, is $\vec{r} = \vec{r}_0$, while the object has initial velocity $\vec{v} = \vec{v}_0$. We seek $\vec{r}(t)$ and $\vec{v}(t)$ for every $t > 0$.

Taking into account that $\vec{a}$ is a constant vector, we have:

$$\frac{d\vec{v}}{dt} = \vec{a} \;\Rightarrow\; d\vec{v} = \vec{a}\,dt \;\Rightarrow\; \int_{\vec{v}_0}^{\vec{v}} d\vec{v} = \vec{a} \int_0^t dt \;\Rightarrow\; \vec{v} - \vec{v}_0 = \vec{a}\,t \;\Rightarrow$$

$$\boxed{\vec{v} = \vec{v}_0 + \vec{a}\,t} \tag{2.17}$$

$$\frac{d\vec{r}}{dt} = \vec{v} = \vec{v}_0 + \vec{a}\,t \;\Rightarrow\; d\vec{r} = \vec{v}_0\,dt + \vec{a}\,t\,dt \;\Rightarrow\; \int_{\vec{r}_0}^{\vec{r}} d\vec{r} = \vec{v}_0 \int_0^t dt + \vec{a} \int_0^t t\,dt \;\Rightarrow$$

$$\boxed{\vec{r} = \vec{r}_0 + \vec{v}_0\,t + \frac{1}{2}\,\vec{a}\,t^2} \tag{2.18}$$

We write:

$$\Delta\vec{r} = \vec{r} - \vec{r}_0 = t\,\vec{v}_0 + \frac{t^2}{2}\,\vec{a}.$$

This vector relation is of the form $\Delta\vec{r} = \kappa\,\vec{v}_0 + \lambda\,\vec{a}$, with constant $\vec{v}_0$, $\vec{a}$ and variable κ, λ. According to Analytic Geometry, the vector $\Delta\vec{r} = \vec{r} - \vec{r}_0$ lies on the constant plane defined by $\vec{v}_0$ and $\vec{a}$ and passing through the point $\vec{r}_0$ (initial position of the moving object). On the same plane will therefore always lie the tip of the position vector $\vec{r}$ of the object. We conclude that

motion with constant acceleration (in magnitude and direction) takes place on a constant plane.[1]

An example is motion of a body in the gravitational field near the surface of the Earth, in a relatively small region of space where this field may be considered uniform. If we ignore air resistance, the body is subject to the constant acceleration of gravity $\vec{g}$ (directed downward, toward the surface of the Earth) and its path is confined to the plane defined by the body's initial velocity $\vec{v}_0$ and the acceleration $\vec{g}$ (the plane of motion is perpendicular to the surface of the Earth).

2.6 Tangential and Normal Components

The velocity $\vec{v}$ and the acceleration $\vec{a}$ of a moving object are vectors of absolute physical and geometrical substance, independent of the choice of a system of axes (x, y, z) in our space. If we choose a different set of axes (x', y', z'), with different origin and orientation relative to (x, y, z), the *components* of $\vec{v}$ and $\vec{a}$ will change but the *vectors* themselves, as geometrical quantities, will remain the same. We will now

---

[1]In the special case where the acceleration is *zero*, the motion is *uniform rectilinear* and the plane is indeterminate.

introduce a system of components that is associated with the trajectory itself of the moving object.

As we already know, a point A of the trajectory can be determined by its position vector $\vec{r}$ relative to any reference point O. This vector is a certain function $\vec{r}(t)$ of time. An alternative way of determining A is the following (see Fig. 2.4). We choose an arbitrary point C of the curve representing the trajectory, as well as a positive direction of motion along the curve (not necessarily coincident with the actual direction of motion). The location of point A on the curve is then given by the distance $s = CA$ of A from C, *measured along the curve*. Note that s may be positive or negative, depending on whether A is located ahead of C or behind C, respectively, in the positive sense of traversing the curve, i.e., in the direction of *increasing s*. To simplify our analysis, we make the assumption that the trajectory is a plane curve. The results we will arrive at, however, will be valid for motion along *any* smooth curve in space.

We consider the points A and A' of the trajectory, through which points the object passes at times t and t', respectively. Let $\vec{r}$ and $\vec{r}'$ be the position vectors of A and A' relative to O. We call $\Delta\vec{r} = \vec{r}' - \vec{r}$ and $\Delta t = t' - t$, and we write:

$$\vec{r} = \vec{r}(t), \quad \vec{r}' = \vec{r}(t') = \vec{r}(t + \Delta t) = \vec{r}(t) + \Delta\vec{r}.$$

The velocity at point A, at time t, is

$$\vec{v} = \frac{d\vec{r}}{dt} = \lim_{\Delta t \to 0} \frac{\vec{r}(t + \Delta t) - \vec{r}(t)}{\Delta t} = \lim_{\Delta t \to 0} \frac{\Delta\vec{r}}{\Delta t}.$$

But, $\frac{\Delta\vec{r}}{\Delta t} = \frac{\Delta\vec{r}}{\Delta s}\frac{\Delta s}{\Delta t}$, so that

$$\lim_{\Delta t \to 0} \frac{\Delta\vec{r}}{\Delta t} = \left(\lim_{\Delta s \to 0} \frac{\Delta\vec{r}}{\Delta s} \right) \left(\lim_{\Delta t \to 0} \frac{\Delta s}{\Delta t} \right)$$

(since $\Delta s \to 0$ when $\Delta t \to 0$). We thus have:

Fig. 2.4 The position A of the moving object is specified by the distance $s = CA$, measured along the trajectory

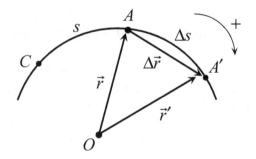

$$\vec{v} = \frac{d\vec{r}}{dt} = \frac{d\vec{r}}{ds}\frac{ds}{dt} \tag{2.19}$$

This relation expresses the derivative of a composite function, given that $\vec{r} = \vec{r}(s)$ and $s = s(t)$, so that $\vec{r} = \vec{r}(t)$.

We now seek the geometrical significance of the vector $d\vec{r}/ds$. We notice that this vector is the limit of $\Delta\vec{r}/\Delta s$ for $\Delta s \to 0$. As $\Delta s \to 0$, the vector $\Delta\vec{r}/\Delta s$ tends to become *tangent* to the trajectory at point A, while its direction is that of *increasing* s ($\Delta s > 0$); that is, it points toward the *positive* direction of traversing the curve. Moreover, $|\Delta\vec{r}/\Delta s| \to 1$ as $\Delta s \to 0$. We conclude that $d\vec{r}/ds$ is a *unit* vector tangent to the trajectory at A and oriented in the *positive* direction of motion on the path. We write:

$$\hat{u}_T = \frac{d\vec{r}}{ds} \tag{2.20}$$

Relation (2.19) now takes on the form

$$\boxed{\vec{v} = \frac{ds}{dt}\hat{u}_T = v\,\hat{u}_T} \tag{2.21}$$

where v is the *algebraic value* of the velocity with respect to the unit vector $\hat{u}_T$:

$$v = \frac{ds}{dt} = \pm|\vec{v}|\,.$$

Thus, if $v > 0$ the motion is in the positive direction (increasing s), while if $v < 0$ the motion is in the negative direction (decreasing s). Note that the unit vector $\hat{u}_T$ is *always* in the *positive* direction, regardless of the direction of motion! (In Fig. 2.5 the motion is in the positive direction; thus $v > 0$.) As seen from (2.21), *the velocity is a vector tangent to the trajectory.*

Our study of acceleration begins with an important remark:

1. A component of acceleration *parallel* to the velocity may alter *the magnitude but not the direction* of the velocity. (This is the case in accelerated rectilinear motion.)

Fig. 2.5 The velocity is a vector tangent to the trajectory at each point of the latter

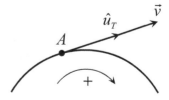

Fig. 2.6 Tangential and
normal acceleration

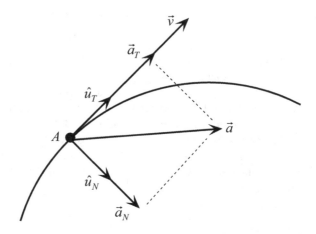

2. A component of acceleration *normal* to the velocity may change *the direction but
    not the magnitude* of the velocity. (This is a direct consequence of the discussion
    in Sect. 2.4.)

Given that the velocity is a vector tangent to the trajectory, it is natural to resolve
the acceleration $\vec{a}$ into two components (see Fig. 2.6):

1. a component $\vec{a}_T$ *tangent* to the trajectory (*tangential acceleration*), which is
    responsible for the change of *speed* (change of *magnitude* of the velocity);
2. a component $\vec{a}_N$ *normal* to the trajectory (*normal* or *centripetal acceleration*),
    which is responsible for the change of *direction* of the velocity.

The total acceleration of the moving object will be the vector sum $\vec{a} = \vec{a}_T + \vec{a}_N$.
To evaluate $\vec{a}_T$ and $\vec{a}_N$ we differentiate (2.21) with respect to t:

$$\vec{a} = \frac{d\vec{v}}{dt} = \frac{d}{dt}(v\,\hat{u}_T) = \frac{dv}{dt}\hat{u}_T + v\frac{d\hat{u}_T}{dt} \tag{2.22}$$

The vector $\frac{d\hat{u}_T}{dt}$ is *normal* to $\hat{u}_T$. Indeed,

$$\hat{u}_T \cdot \frac{d\hat{u}_T}{dt} = \frac{1}{2}\frac{d}{dt}(\hat{u}_T \cdot \hat{u}_T) = \frac{1}{2}\frac{d}{dt}(|\hat{u}_T|^2) = 0\,,$$

given that $|\hat{u}_T| = 1 = $ constant. Moreover, $\frac{d\hat{u}_T}{dt}$ is directed toward the *concave*
("inner") side of the trajectory, since this is the case with the infinitesimal change
$d\hat{u}_T$ of $\hat{u}_T$. [By (2.22), then, *the acceleration $\vec{a}$ of the moving object is oriented toward
the concavity of the curve.*] We thus consider a unit vector $\hat{u}_N$ normal to the trajectory
and directed "inward", and we write:

$$\frac{d\hat{u}_T}{dt} = \left|\frac{d\hat{u}_T}{dt}\right|\hat{u}_N = \frac{|d\hat{u}_T|}{dt}\,\hat{u}_N \tag{2.23}$$

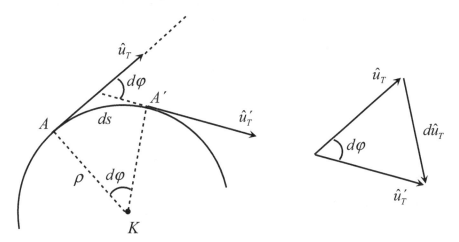

Fig. 2.7 The infinitesimal section ds of the trajectory may be regarded as an arc of a local circle with center K and radius ρ, these quantities generally varying along the trajectory. (The trajectory itself need not be a circle macroscopically!)

The differential $d\hat{u}_T$ is approximately equal to an infinitesimal change of $\hat{u}_T$ when this unit vector is displaced along the curve within an infinitesimal time interval dt. That is, $d\hat{u}_T = \hat{u}'_T - \hat{u}_T$, as seen in Fig. 2.7.

The $\hat{u}_T$ and $\hat{u}'_T$ are unit vectors tangent to the curve at the respective points A and A'. These points are an infinitesimal distance ds apart (measured *along* the curve), which distance the object covers within time dt. Given that the angle $d\varphi$ between $\hat{u}_T$ and $\hat{u}'_T$ is infinitesimal, we may write:

$$|d\hat{u}_T| \simeq |\hat{u}_T| d\varphi = d\varphi , \quad \text{where } d\varphi \text{ is in } rad .$$

Relation (2.23) is thus written (by taking into account that $v = ds/dt$):

$$\frac{d\hat{u}_T}{dt} = \frac{d\varphi}{dt} \hat{u}_N = \frac{d\varphi}{ds} \frac{ds}{dt} \hat{u}_N = \frac{d\varphi}{ds} v \hat{u}_N \qquad (2.24)$$

Given that ds is infinitesimal, it may approximately be regarded as an arc of a local circle with center K (the point of intersection of the lines normal to the curve at A and A') and radius $\rho = AK$ (Fig. 2.7). The point K is called the *center of curvature* of the trajectory at A, while ρ represents the *radius of curvature* at A (these quantities generally vary along the trajectory). Thus, $ds \simeq \rho d\varphi$, so that (2.24) becomes:

$$\frac{d\hat{u}_T}{dt} = \frac{v}{\rho} \hat{u}_N \tag{2.25}$$

Finally, (2.22) yields:

$$\vec{a} = \frac{dv}{dt} \hat{u}_T + \frac{v^2}{\rho} \hat{u}_N \tag{2.26}$$

We write:

$$\vec{a} = a_T \hat{u}_T + a_N \hat{u}_N$$
$$a_T = \frac{dv}{dt} = \frac{d^2 s}{dt^2} , \qquad a_N = \frac{v^2}{\rho} \tag{2.27}$$

The magnitude of the acceleration is

$$|\vec{a}| = \sqrt{a_T^2 + a_N^2} = \left[\left(\frac{dv}{dt}\right)^2 + \frac{v^4}{\rho^2} \right]^{1/2} \tag{2.28}$$

We stress that these results are valid for *any curve in space*, not just for motion on a plane curve.

Let us summarize the ways of resolving velocity and acceleration into components:

$$\vec{v} = v_x \hat{u}_x + v_y \hat{u}_y + v_z \hat{u}_z = v \hat{u}_T$$
$$\vec{a} = a_x \hat{u}_x + a_y \hat{u}_y + a_z \hat{u}_z = a_T \hat{u}_T + a_N \hat{u}_N \tag{2.29}$$

Note, in particular, that

$$|\vec{a}|^2 = a_x^2 + a_y^2 + a_z^2 = a_T^2 + a_N^2 \tag{2.30}$$

Special cases:

1. *Uniform curvilinear motion: v = constant.* We have:

$$a_T = \frac{dv}{dt} = 0 , \qquad \vec{a} = a_N \hat{u}_N = \frac{v^2}{\rho} \hat{u}_N \tag{2.31}$$

Note that in uniform motion the acceleration is normal to the velocity, in accordance with the conclusions of Sect. 2.4. [*Exercise*: Show that the distance s along the curve (Fig. 2.4) is given as a function of time by $s = vt$, where we assume that $s = 0$ at $t = 0$.]

2. *Rectilinear motion:* $\rho = \infty$, $s = x$, $\hat{u}_T = \hat{u}_x$. Hence,

$$\vec{v} = \frac{ds}{dt}\,\hat{u}_T = \frac{dx}{dt}\,\hat{u}_x = v\,\hat{u}_x\,,$$

$$a_N = \frac{v^2}{\rho} = 0\,, \quad \vec{a} = a_T\,\hat{u}_T = \frac{dv}{dt}\,\hat{u}_x\,.$$

In *uniform rectilinear* motion ($\vec{v} = const.$) both a_T and a_N are zero, so that $\vec{a} = 0$.

2.7 Circular Motion

Circular motion is plane motion with constant radius of curvature $\rho = R$. The trajectory of the moving object is a circle of radius R (Fig. 2.8).

We choose the positive direction of motion to be counterclockwise, in which case the length s of the arc from a reference point C, as well as the angle θ measured from C, increase counterclockwise and decrease clockwise (Fig. 2.8 represents motion in the positive direction). As always, the unit tangent vector $\hat{u}_T$ is oriented in the positive direction regardless of the actual direction of motion.

As we know, $s = R\theta$, where θ is measured in *rad*. The velocity of the object is

$$\vec{v} = v\,\hat{u}_T \quad \text{where} \quad v = \frac{ds}{dt} = R\frac{d\theta}{dt} \tag{2.32}$$

(In general, $v = \pm|\vec{v}|$, depending on the direction of motion.) We define the *angular velocity*

$$\omega = \frac{d\theta}{dt} \tag{2.33}$$

Fig. 2.8 Circular motion is plane motion with constant radius of curvature $\rho = R$

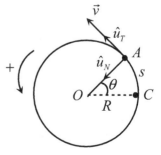

Then,

$$\boxed{v = R\omega}$$ (2.34)

The quantity ω is measured in $rad/s = rad.s^{-1}$. Note that the sign of ω is the same as that of v and depends on the direction of motion.

The acceleration of the object is

$$\vec{a} = a_T \, \hat{u}_T + a_N \, \hat{u}_N$$

where

$$a_T = \frac{dv}{dt} = R\frac{d\omega}{dt} \quad , \quad a_N = \frac{v^2}{R} = \frac{(R\omega)^2}{R}.$$

We define the *angular acceleration*

$$\alpha = \frac{d\omega}{dt}$$ (2.35)

We thus have:

$$\boxed{a_T = R\alpha \, , \quad a_N = R\omega^2}$$ (2.36)

The magnitude of the acceleration is

$$|\vec{a}| = \sqrt{a_T^2 + a_N^2} = R\sqrt{\alpha^2 + \omega^4}$$ (2.37)

In *uniform circular motion* the algebraic values v and ω are constant, so that, by (2.35) and (2.36), $\alpha = 0$ and $a_T = 0$. Hence, the acceleration is purely *centripetal*:

$$\vec{a} = \frac{v^2}{R} \, \hat{u}_N = R\omega^2 \, \hat{u}_N \quad (v, \omega = const.)$$ (2.38)

By (2.33) and by taking into account that ω is constant, we have (assuming that $\theta = \theta_0$ at $t = 0$):

$$d\theta = \omega \, dt \quad \Rightarrow \quad \int_{\theta_0}^{\theta} d\theta = \int_0^t \omega \, dt = \omega \int_0^t dt \quad \Rightarrow$$

$$\theta = \theta_0 + \omega \, t$$ (2.39)

Exercise: Show that the equation for the arc length s is $s = s_0 + vt$.

The *period T* (measured in s) and the *frequency* $f = 1/T$ (measured in s^{-1} or *hertz*, *Hz*) of uniform circular motion are defined by the relation

$$|\omega| = \frac{2\pi}{T} = 2\pi f \tag{2.40}$$

More on these will be said in Chap. 5, in the context of simple harmonic motion.

2.8 Relative Motion

Consider two objects A, B and an observer O who uses the coordinate system (x, y, z). This system is called the *frame of reference* of O. The position vectors of A and B relative to O are $\vec{r}_A$ and $\vec{r}_B$, respectively, while the velocities of A and B with respect to O are

$$\vec{v}_A = \frac{d\vec{r}_A}{dt} , \quad \vec{v}_B = \frac{d\vec{r}_B}{dt}.$$

We call $\vec{r}_{BA} = \vec{r}_B - \vec{r}_A$ the position vector of B relative to A (Fig. 2.9). The velocity of B *relative to A* is defined as

$$\vec{v}_{BA} = \frac{d\vec{r}_{BA}}{dt} \tag{2.41}$$

We notice that

$$\vec{v}_{BA} = \frac{d}{dt}(\vec{r}_B - \vec{r}_A) = \frac{d\vec{r}_B}{dt} - \frac{d\vec{r}_A}{dt} \quad \Rightarrow$$

$$\vec{v}_{BA} = \vec{v}_B - \vec{v}_A \tag{2.42}$$

Similarly, the velocity of A relative to B is

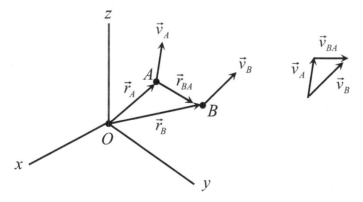

Fig. 2.9 Relative motion of two objects A and B

$$\vec{v}_{AB} = \vec{v}_A - \vec{v}_B = -\vec{v}_{BA} \tag{2.43}$$

In an analogous way we define the acceleration of B relative to A:

$$\vec{a}_{BA} = \frac{d\vec{v}_{BA}}{dt} = \frac{d}{dt}(\vec{v}_B - \vec{v}_A) = \frac{d\vec{v}_B}{dt} - \frac{d\vec{v}_A}{dt} \quad \Rightarrow$$

$$\vec{a}_{BA} = \vec{a}_B - \vec{a}_A = -\vec{a}_{AB} \tag{2.44}$$

where

$$\vec{a}_A = \frac{d\vec{v}_A}{dt} , \qquad \vec{a}_B = \frac{d\vec{v}_B}{dt}$$

are the accelerations of A and B with respect to O.

Consider now two observers O and O', where O' is moving with *constant* velocity $\vec{V}$ relative to O; that is, $\vec{v}_{O'O} = \vec{V} = const$. The relative acceleration of these observers is, therefore, zero:

$$\vec{a}_{O'O} = \frac{d\vec{v}_{O'O}}{dt} = 0.$$

Assume, further, that these observers record the motion of a particle Σ (Fig. 2.10). We denote the velocity and the acceleration of Σ with respect to the two observers as follows: $\vec{v}_{\Sigma O} = \vec{v}$, $\vec{a}_{\Sigma O} = \vec{a}$; $\vec{v}_{\Sigma O'} = \vec{v}'$, $\vec{a}_{\Sigma O'} = \vec{a}'$.

We now apply (2.42) and (2.44) with O' in place of A and Σ in place of B:

$$\vec{v}_{\Sigma O'} = \vec{v}_{\Sigma O} - \vec{v}_{O'O} \quad \Rightarrow \quad \vec{v}' = \vec{v} - \vec{V} ;$$
$$\vec{a}_{\Sigma O'} = \vec{a}_{\Sigma O} - \vec{a}_{O'O} \quad \Rightarrow \quad \vec{a}' = \vec{a} .$$

We thus arrive at an important conclusion:

A particle moves with the same acceleration with respect to two observers that maintain a constant velocity relative to each other (they do not accelerate with respect to each other).

Fig. 2.10 The motion of a particle Σ as seen by two observers O and O' moving with constant velocity relative to each other

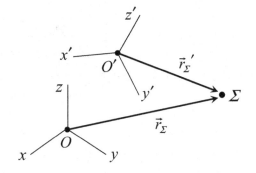

In particular,

if the particle moves with constant velocity relative to one observer, it will also move with constant velocity relative to the other observer.

Stated differently, if the particle executes *uniform rectilinear* motion relative to one observer, it will execute the same kind of motion relative to the other observer.

References

1. A.F. Bermant, I.G. Aramanovich, *Mathematical Analysis* (Mir Publishers, 1975)
2. D.D. Berkey, *Calculus*, 2nd edn. (Saunders College, 1988)

Chapter 3
Dynamics of a Particle

Abstract Newton's laws are presented and the concept of inertial reference frames is defined. Gravitational and frictional forces are studied. The concepts of angular momentum and torque are introduced, and the properties of central forces are examined.

3.1 The Law of Inertia

The term *point particle* (or simply *particle*) refers to a body whose dimensions are so small that we may ignore its rotational motion (if any). But, even a body of *finite* dimensions may be treated as a "particle" if its motion is *purely translational* (that is, if the body is not rotating).

A particle is said to be *free* if (*a*) it is not subject to any interactions with the rest of the world (a case that is rather unrealistic!) or (*b*) the totality of its interactions somehow sum to zero (i.e., they cancel one another) so that the particle behaves *as if* it were subject to no interactions whatsoever. According to the *Law of Inertia* or *Newton's First Law*,

a free particle moves with constant velocity (i.e., has no acceleration) relative to any other free particle.

Therefore, a free particle either is in *uniform rectilinear* motion, or is at rest, relative to another free particle.

Imagine now an observer who herself is a free particle (this is approximately true for someone who is at rest on the surface of the Earth). Such an observer is called an *inertial observer* and the system of coordinates or axes she uses is called an *inertial frame of reference* (or, simply, *inertial frame*). According to the law of inertia,

different inertial observers move with constant velocities (thus, do not accelerate) relative to one another.

For example, the passenger in a train that moves with constant velocity relative to the ground is (approximately) an inertial observer and a fixed system of axes (x, y, z) in the train is an inertial frame of reference.

C. J. Papachristou, *Introduction to Mechanics of Particles and Systems*,
https://doi.org/10.1007/978-3-030-54271-9_3

On the basis of the law of inertia we may now give the following definition of an inertial reference frame:

> An inertial frame of reference is any set of coordinates (or axes) relative to which a free particle either moves with constant velocity or is at rest. Thus, in an inertial frame a free particle does not accelerate.

We note that the observer who uses this frame is, by definition, *at rest* relative to it.

A reference frame that *accelerates* with respect to an inertial frame is obviously *not* inertial. This is, e.g., the case with the Earth because of its daily rotation as well as its orbiting motion about the Sun (if we regard the latter as an almost inertial frame). However, since the acceleration of the Earth is relatively small compared to the accelerations measured in typical terrestrial experiments, we may, for practical purposes, consider the Earth as an almost inertial frame. Hence, *any observer moving on the surface of the Earth with constant velocity will be regarded as an inertial observer.*

3.2   Momentum, Force, and Newton's 2nd and 3rd Laws

In an abstract sense, *force* represents the effort necessary in order to alter the state of motion of a body; in particular, in order to change the body's velocity. The first idea that comes to mind is to quantitatively identify force with acceleration. We know from our experience, however, that different bodies generally require a different effort in order to acquire the same acceleration, or, the same velocity within the same period of time. (Try, e.g., to produce the same acceleration on a book and on a truck by pushing them!) This happens because different bodies exhibit different *inertia*, that is, different resistance to a change of their state of motion. This property must therefore be taken into account in the definition of force. To this end, we introduce a new physical quantity called *linear momentum* (or simply *momentum*) of a body:

$$\vec{p} = m\vec{v} \tag{3.1}$$

where $\vec{v}$ is the velocity of the body. The coefficient m is called *mass* and is a measure of the body's inertia.

Newton's Second Law of Motion (we will often simply call it "Newton's Law"), which is valid only in *inertial* frames of reference, in essence *defines* the force exerted on a body as the rate of change of the body's momentum at time t:

$$\boxed{\vec{F} = \frac{d\vec{p}}{dt}} \tag{3.2}$$

If we make the assumption that the mass m is constant, then

$$\frac{d\vec{p}}{dt} = \frac{d}{dt}(m\vec{v}) = m\frac{d\vec{v}}{dt}.$$

Hence,

$$\boxed{\vec{F} = m\,\vec{a}} \tag{3.3}$$

where $\vec{a} = d\vec{v}/dt$ is the acceleration of the body at time t. We stress that the form (3.3) of Newton's law is valid for a body of constant mass, as well as that relations (3.2) and (3.3) are valid on the assumption that the observer measuring the velocity and the acceleration of the body is an *inertial* observer. By using (3.3) and by taking into account the conclusion at the end of Sect. 2.8, it is not hard to show that

the force on a particle is the same for all inertial observers

(recall that inertial observers move with constant velocities relative to one another).

The vector Eq. (3.3) is equivalent to three algebraic equations, one for each vector component. We write:

$$\vec{F} = F_x\hat{u}_x + F_y\hat{u}_y + F_z\hat{u}_z , \quad \vec{a} = a_x\hat{u}_x + a_y\hat{u}_y + a_z\hat{u}_z .$$

Substituting these expressions into (3.3) and equating corresponding components in the two sides of this equation, in accordance with (1.8), we have:

$$F_x = ma_x , \quad F_y = ma_y , \quad F_z = ma_z \tag{3.4}$$

Now, as implied by the law of inertia, the change of the state of motion of a body *relative to an inertial observer* requires an interaction of the body with the rest of the world. The force $\vec{F}$ is precisely a measure of this interaction. If the body is not subject to interactions (i.e., is a free "particle") then $\vec{F} = 0$ and it follows from (3.3) that the velocity of the body with respect to an *inertial* reference frame is constant (since the acceleration is zero). We thus conclude that the second law of motion is consistent with the law of inertia, provided that both these laws are examined from the point of view of an inertial frame of reference.

It is tempting to argue that, according to the above discussion, the law of inertia is redundant since *it appears* to be just a special case of the second law:

Free particle ⟺ no interaction ⟺ no force ⟺ no acceleration ⟺ constant velocity.

There is a subtle point, however: What kind of observer is *entitled* to conclude that a particle that appears to move with constant velocity (i.e., with no acceleration) is a free particle? *Answer:* Only an *inertial* observer, who uses an *inertial* frame of reference! The purpose of the law of inertia is essentially to *define* these frames and *guarantee* their existence. So, without the first law of motion, the second law would become indeterminate, if not altogether wrong, since it would appear to be valid relative to any observer regardless of his or her state of motion. One may say that the first law defines the "terrain" within which the second law acquires a meaning. Applying the latter law without taking the former one into account would be like trying to play soccer without possessing a soccer field!

According to (3.2), if a body is not subject to any force ($\vec{F} = 0$) its momentum $\vec{p}$ relative to an inertial frame is constant in time, since, in this case, $d\vec{p}/dt = 0$. As will be seen in Chap. 6, this is true, more generally, for any *isolated* system of particles, i.e., a system subject to no *external* interactions. For such a system the *principle of conservation of momentum* is valid:

The total momentum of a system of particles subject to no external forces is constant in time.

This principle is intimately related to a third law of motion. Consider a system of two particles subject only to their mutual interaction (there are no *external* forces). The total momentum of the system at times t and $t + \Delta t$ is

$$\vec{P}(t) = \vec{p}_1 + \vec{p}_2 = m_1\vec{v}_1 + m_2\vec{v}_2 \ ,$$
$$\vec{P}(t + \Delta t) = \vec{p}_1' + \vec{p}_2' = m_1\vec{v}_1' + m_2\vec{v}_2' \ .$$

By conservation of momentum,

$$\vec{P}(t) = \vec{P}(t + \Delta t) \ \Rightarrow \ \vec{p}_1 + \vec{p}_2 = \vec{p}_1' + \vec{p}_2' \ \Rightarrow$$

$$\vec{p}_1' - \vec{p}_1 = -(\vec{p}_2' - \vec{p}_2) \ \Leftrightarrow \ \Delta\vec{p}_1 = -\Delta\vec{p}_2 \tag{3.5}$$

Hence,

$$\frac{\Delta\vec{p}_1}{\Delta t} = -\frac{\Delta\vec{p}_2}{\Delta t} \ \Rightarrow \ \lim_{\Delta t \to 0}\frac{\Delta\vec{p}_1}{\Delta t} = -\lim_{\Delta t \to 0}\frac{\Delta\vec{p}_2}{\Delta t} \ \Rightarrow \ \frac{d\vec{p}_1}{dt} = -\frac{d\vec{p}_2}{dt} \ .$$

But, by (3.2),

$$\frac{d\vec{p}_1}{dt} = \vec{F}_{12} \ , \qquad \frac{d\vec{p}_2}{dt} = \vec{F}_{21}$$

where $\vec{F}_{12}$ is the *internal* force exerted on particle *1* by its interaction with particle 2, while $\vec{F}_{21}$ is the force on particle 2 due to its interaction with particle *1*. Thus, finally,

$$\boxed{\vec{F}_{12} = -\vec{F}_{21}} \tag{3.6}$$

Relation (3.6) expresses *Newton's Third Law* or *Law of Action and Reaction*. Note that this law is equivalent to the principle of conservation of momentum, which principle, in turn, constitutes the generalization of the law of inertia for a system of particles. Taking into account that Newton's second law (in essence, the *definition* of the concept of force) also is a logical extension of the first law, we can appreciate the great importance of the law of inertia in the axiomatic foundation of Newtonian Mechanics! (For a discussion of the axiomatic basis of Newtonian Mechanics, see [1].)

You may have noticed that we defined momentum, which depends explicitly on mass, without previously giving a definition of mass itself. We will now describe a method for determining mass, based on the principle of conservation of momentum. Consider again an *isolated* system (i.e., a system subject to no external forces) consisting of two particles of masses m_1, m_2, which somehow interact with each other (e.g., they collide or, if they are electrically charged, they exert Coulomb forces on each other, etc.). Assume that, within a time interval Δt, the momenta of m_1 and m_2 change by $\Delta \vec{p}_1$ and $\Delta \vec{p}_2$, respectively. According to (3.5), $\Delta \vec{p}_1 = -\Delta \vec{p}_2$ or

$$m_1 \Delta \vec{v}_1 = -m_2 \Delta \vec{v}_2 \tag{3.7}$$

In terms of magnitudes,

$$m_1 |\Delta \vec{v}_1| = m_2 |\Delta \vec{v}_2| \quad \Rightarrow$$

$$\frac{m_2}{m_1} = \frac{|\Delta \vec{v}_1|}{|\Delta \vec{v}_2|} \tag{3.8}$$

As experiment shows, the ratio of magnitudes on the right-hand side of (3.8) always assumes the same value for given particles m_1 and m_2, regardless of the kind or the duration of their interaction. Moreover, the vectors $\Delta \vec{v}_1$ and $\Delta \vec{v}_2$ are found to be in opposite directions, in accordance with (3.7). These observations verify that to each particle in the system there corresponds a constant quantity m, its mass, such that the sum $m_1 \vec{v}_1 + m_2 \vec{v}_2$ retains a constant value when the particles are subject only to their mutual interaction. This constitutes an experimental verification of conservation of momentum.

Now, relation (3.8) allows us to determine the ratio m_2/m_1 experimentally by measuring the ratio of magnitudes of $\Delta \vec{v}_1$ and $\Delta \vec{v}_2$. Hence, by *arbitrarily* assigning unit value to the mass of particle *1*, we can determine the mass of particle *2* as follows: We allow the two particles to interact for a time interval Δt; then we measure the (vector) changes of their velocities within this interval and we calculate the ratio of magnitudes of these changes. The result yields the mass m_2 of particle *2* numerically (since, by definition, m_1 is a unit mass). In a similar way, we determine the mass m of any particle by allowing this particle to interact with a particle of known mass. By measuring the instantaneous acceleration $\vec{a}$ of m we then find the corresponding instantaneous force $\vec{F}$ on this particle by using Newton's law (3.3).

In the S.I. system of units, the unit of mass is $1\ kg = 10^3\ g$ while the unit of force is $1\ Newton = 1\ N = 1\ kg.m.s^{-2}$.

Assume now that a body of mass m is subject to various interactions with its surroundings, which interactions are represented quantitatively by the forces $\vec{F}_1$, $\vec{F}_2$, ... The vector sum

$$\sum_i \vec{F_i} \equiv \sum \vec{F} = \vec{F_1} + \vec{F_2} + \cdots$$

is the *resultant force* (or *total force*) acting on the body. Newton's second law then takes on the form:

$$\boxed{\sum \vec{F} = \frac{d\vec{p}}{dt} = m\,\vec{a}} \tag{3.9}$$

Let a_x, a_y, a_z be the components of the acceleration $\vec{a}$ of the body. According to (1.10), the components of the resultant force $\sum \vec{F}$ are ΣF_x, ΣF_y, ΣF_z, where ΣF_x is the sum of the x-components of the individual forces $\vec{F_1}$, $\vec{F_2}$, ..., and similarly for ΣF_y and ΣF_z. By equating corresponding components on the two sides of (3.9), we have:

$$\boxed{\sum F_x = ma_x, \quad \sum F_y = ma_y, \quad \sum F_z = ma_z} \tag{3.10}$$

As an example, assume that a body of mass $m = 2\ kg$ is subject to the forces $\vec{F_1} \equiv (3, 1, -1)N$ and $\vec{F_2} \equiv (-1, 3, -1)N$. The resultant force is

$$\sum \vec{F} = \vec{F_1} + \vec{F_2} \equiv (2, 4, -2)N$$

so that $\Sigma F_x = 2\,N$, $\Sigma F_y = 4\,N$, $\Sigma F_z = $ -2 N. By (3.10) we find the acceleration of the body: $\vec{a} \equiv (a_x, a_y, a_z) \equiv (1, 2, -1)\ m \cdot s^{-2}$.

A body is said to be in *equilibrium* if the total force on it is zero: $\Sigma \vec{F} = 0$. Note that by "equilibrium" we do not necessarily mean *rest* ($\vec{v} = 0$). According to (3.9), a body is in a state of equilibrium if it does not accelerate ($\vec{a} = 0$) relative to an inertial observer. If, however, the body is initially at rest at an equilibrium position where the total force on it vanishes, then it remains at rest there.

Conversely, a body may be *momentarily* at rest without being in a state of equilibrium. The total force acting on it will then cause an acceleration that will put the body back in motion at the very next moment. For example, if we throw a stone straight upward, it will stop instantaneously as soon as it reaches a maximum height and then it will start moving downward immediately, under the action of gravity. Another example of momentary rest is that of a pendulum bob at the highest point of its path.

We noted earlier that a body of finite dimensions can be treated as if it were a point particle if its motion is purely translational (i.e., the body is not subject to rotation). Such a motion depends only on the resultant force on the body, regardless of the location of the points of action of the various individual forces that act on this body. On the contrary, as will be seen in Chap. 7, the points of action of these forces *are* important when one considers rotational motion, as this motion is determined by the total external *torque* on the body.

3.3 Force of Gravity

Near the surface of the Earth and in the absence of air resistance, all bodies fall toward the ground with a common acceleration $\vec{g}$, called the *acceleration of gravity* and having a magnitude $g \simeq 9.8 \ m \cdot s^{-2}$. The force of gravitational attraction between a body and the Earth is called the *weight* $\vec{w}$ of the body. If m is the mass of the body, then, by Newton's second law,

$$\vec{w} = m\vec{g} \tag{3.11}$$

For larger distances from the surface of the Earth, the value of g (hence also the weight of a body) varies as a function of the distance from the Earth. We call M and R the mass and the radius of the Earth, respectively, and we let h be a given height above the surface of the Earth. We would like to determine the value of g at this height. According to *Newton's Law of Gravity*, the magnitude of the gravitational force on a body of mass m, located at a height h above the Earth, is

$$w = G\frac{Mm}{(R+h)^2} \tag{3.12}$$

where G is a constant, the value of which is experimentally determined to be

$$G = 6.673 \times 10^{-11} \mathrm{N} \cdot \mathrm{m}^2 \cdot \mathrm{kg}^{-2}.$$

Taking into account that $w = mg$, we find that

$$g = \frac{GM}{(R+h)^2} \tag{3.13}$$

Note that the ratio w/m, which represents the *gravitational field strength* at the considered location, also represents the acceleration of gravity, g. According to (3.13), this acceleration is independent of the mass m of the body. Thus, *all bodies experience the same acceleration at any point in a gravitational field*, regardless of the particular physical properties of each body (provided, of course, that no forces other than gravity are present).

3.4 Frictional Forces

Sliding friction (or simply *friction*) is a force that tends to oppose the relative motion of two surfaces when they are in contact. It is a cumulative effect of a large number of microscopic interactions of electromagnetic origin, among the atoms or molecules of the two surfaces. Practically speaking, these surfaces belong to two bodies that

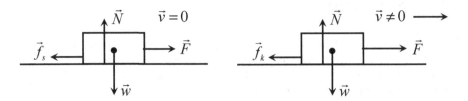

Fig. 3.1 A box pushed to the right experiences static (left figure) or kinetic (right figure) friction, depending on whether it is at rest or in motion, respectively

are in contact with each other (although true contact is never possible at the atomic level!).

Let us consider, for example, a box of weight $\vec{w}$ lying on the horizontal surface of a table (Fig. 3.1). The box is initially at rest under the action of two forces, namely, its weight and the normal reaction $\vec{N}$ from the table. This state of equilibrium implies that the resultant force on the box is zero: $\vec{N} + \vec{w} = 0 \Leftrightarrow \vec{N} = -\vec{w}$.

We now push the box to the right with a force $\vec{F}$ that may vary in magnitude. The box "wants" to slide to the right but there is a force $\vec{f}$ opposing this motion. This force, which is directed to the left, is the friction between the box and the surface of the table. If $\vec{F}$ is not large enough, $\vec{f}$ manages to balance it and the box remains at rest. We say that $\vec{f}$ is *static friction* and we denote it by $\vec{f}_s$. Obviously, $\vec{F} + \vec{f}_s = 0$. Depending on the applied force $\vec{F}$, the force $\vec{f}_s$ varies from zero (when $\vec{F} = 0$) to a maximum value $\vec{f}_{s,\max}$.

When $\vec{F}$ exceeds $f_{s,\max}$ in magnitude, the box is set in motion on the table, accelerating to the right. The frictional force then decreases from $f_{s,\max}$ to a new, constant value $\vec{f}_k$ (also directed to the left) that opposes the motion; it is called *kinetic friction*. The total force on the box during the motion is $\vec{F}_{tot} = \vec{F} + \vec{f}_k$. If we assume that the box moves in the positive direction of the x-axis, then

$$\vec{F} = |\vec{F}|\hat{u}_x = F\hat{u}_x, \quad \vec{f}_k = -\left|\vec{f}_k\right|\hat{u}_x = -f_k\hat{u}_x$$

so that

$$\vec{F}_{tot} = F\,\hat{u}_x + (-f_k)\,\hat{u}_x = (F - f_k)\,\hat{u}_x \qquad (3.14)$$

By dividing $\vec{F}_{tot}$ with the mass m of the box we find the acceleration $\vec{a}$ of the box. In the case where $F = f_k$ the resultant force $\vec{F}_{tot}$ is zero and the box moves with constant velocity. If we withdraw the applied force $\vec{F}$, the kinetic friction $\vec{f}_k$ decelerates the body until the latter comes to a halt.

As found experimentally, the magnitude $f_{s,\max}$ of *maximum* static friction, as well as the magnitude f_k of kinetic friction, is proportional to the magnitude N of the normal force pressing one surface against the other. Thus, the possible values of static friction are

Fig. 3.2 Experimental
setting for measuring the
coefficients of friction

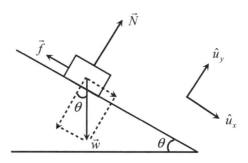

$$0 \le f_s \le f_{s,\max} = \mu_s N \qquad (3.15)$$

while the value of kinetic friction is

$$f_k = \mu_k N \qquad (3.16)$$

where μ_s and μ_k are the *coefficients of friction* (static and kinetic, respectively), with $\mu_k < \mu_s$. Note that μ_s and μ_k are *dimensionless* quantities (i.e., pure numbers). These coefficients depend on the nature of the two surfaces and typically vary between 0.05 and 1.5.

We now describe an experimental method for determining the coefficients of friction between two surfaces:

Consider an inclined plane of variable angle θ, on which plane we have placed a box of mass m (Fig. 3.2). For small values of θ the box is at rest, since the static friction f_s balances the component $w_x = mg \sin \theta$ of the weight along the plane. By gradually increasing the angle θ, we notice that the box remains at rest until this angle exceeds a certain value $\theta = \theta_c$, after which the body begins to slide on the plane.

The box is subject to three forces; namely, its weight $\vec{w} = m\vec{g}$, the normal force $\vec{N}$ from the plane, and the friction $\vec{f}$ (static $\vec{f}_s$ or kinetic $\vec{f}_k$, depending on whether $\vec{v} = 0$ or $\vec{v} \neq 0$, respectively). It is convenient to resolve the weight into two mutually perpendicular components: $w_x = mg \sin \theta$, along the plane, and $w_y = mg \cos \theta$ normal to it. Depending on the value of θ, the following two physical situations are possible:

a. For $\theta \le \theta_c$, the body stays at rest. By the condition for equilibrium,

$$\sum \vec{F} = \vec{w} + \vec{N} + \vec{f}_s = 0$$

or, in terms of components,

$$\sum F_x = 0 \Rightarrow mg \sin \theta - f_s = 0 \Rightarrow mg \sin \theta = f_s,$$
$$\sum F_y = 0 \Rightarrow N - mg \cos \theta = 0 \Rightarrow mg \cos \theta = N.$$

By dividing the equations on the right, we get:

$$\tan\theta = \frac{f_s}{N} \Rightarrow f_s = N\tan\theta.$$

But,

$$f_s \leq f_{s,\max} \Rightarrow N\tan\theta \leq \mu_s N \Rightarrow \tan\theta \leq \mu_s.$$

In the limit case $\theta = \theta_c$ we have $f_s = f_{s,\max}$ and

$$\tan\theta_c = \mu_s \qquad\qquad (3.17)$$

By measuring the angle θ_c we determine the coefficient μ_s.

b. For $\theta > \theta_c$ the body accelerates along the inclined plane; the friction is now kinetic. If we gradually decrease the angle θ we will find some value $\theta_c' < \theta_c$ for which the body moves with constant velocity. By Newton's law and by taking into account that the body does not accelerate, we have:

$$\sum \vec{F} = \vec{w} + \vec{N} + \vec{f_k} = 0.$$

Taking components, as before, we find:

$$mg\sin\theta_c' = f_k = \mu_k N, \quad mg\cos\theta_c' = N.$$

By dividing these equations, we get:

$$\tan\theta_c' = \mu_k \qquad\qquad (3.18)$$

The experimental fact that $\theta_c' < \theta_c$ combined with (3.17) and (3.18) indicates that $\mu_k < \mu_s$. This means that $f_k < f_{s,\max}$.

3.5 Systems with Variable Mass

In the case of a point particle or a body of constant mass m, Newton's second law may be expressed in two equivalent ways:

$$\vec{F} = \frac{d\vec{p}}{dt}\,(a) \Leftrightarrow \vec{F} = m\vec{a}\,(b)$$

However, in the case of a *system* of particles relation (b) is not applicable, since it is not clear what exactly the acceleration vector represents (in Chap. 6 we will see that this relation acquires a meaning by introducing the concept of the center of

Fig. 3.3 A moving platform
on which sand falls at a
constant rate

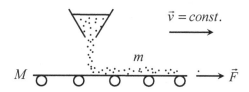

mass). Thus, in the mechanics of systems we generally use relation (a), in the form

$$\vec{F} = \frac{d\vec{P}}{dt} \qquad (3.19)$$

where $\vec{P}$ is the *total* momentum of the system at time t, and where $\vec{F}$ is the total
external force acting on the system at this instant (we will prove this equation in
Chap. 6).

There are systems whose *parts* have variable masses due to a redistribution of the
total mass of the system (which mass is assumed to be *constant* for the time interval
of interest). As an example, consider a moving platform on which sand falls at a
rate of α *kg/s* (Fig. 3.3). We want to find the force $\vec{F}$ with which we must pull the
platform in order for it to move with a constant velocity $\vec{v}$.

Let M be the (constant) mass of the platform. We call m the mass of the sand that
has already fallen on the platform at time t, and we let dm be the additional mass of
sand that falls within an infinitesimal time interval dt. According to the data of the
problem,

$$\frac{dm}{dt} = \alpha \quad (kg/s) \qquad (3.20)$$

Now, to use relation (3.19) we must first decide to which system of masses it will
be applied. The *total* mass of this system will be *constant* in the time interval dt,
although this mass will suffer redistribution. As a "system" we consider the three
masses M, m and dm. At time t the mass $(M + m)$ moves with velocity $\vec{v}$ while the
velocity of dm is zero (this latter quantity has not yet fallen onto the platform). At
time $t + dt$, however, the total mass $(M + m+dm)$ moves with velocity $\vec{v}$. If $\vec{P}(t)$ and
$\vec{P}(t + dt)$ is the total momentum of the system at these two instants, we have:

$$\vec{P}(t) = (M + m)\vec{v} + (dm) \cdot 0, \quad \vec{P}(t + dt) = (M + m + dm)\vec{v}.$$

The change of the system's momentum within the time interval dt is

$$d\vec{P} = \vec{P}(t + dt) - \vec{P}(t) = (dm)\vec{v} \quad \Rightarrow \quad \frac{d\vec{P}}{dt} = \frac{dm}{dt}\vec{v} = \alpha\vec{v}$$

where we have used (3.20). According to (3.19), $d\vec{P}/dt$ represents the total external force on the system at time t, which force is none other than the force $\vec{F}$ we apply on the platform. Hence,

$$\vec{F} = \frac{dm}{dt}\vec{v} = \alpha\vec{v} \qquad (3.21)$$

Notice that the force in (3.21) is proportional to the velocity rather than to the acceleration (which here is zero)!

3.6 Tangential and Normal Components of Force

Recall from Sect. 2.6 that, in curvilinear motion the acceleration $\vec{a}$ is the vector sum of a tangential component $\vec{a}_T$ (tangent, that is, to the trajectory of the moving particle) and a normal (or centripetal) component $\vec{a}_N$, as seen in Fig. 3.4.

We write:

$$\vec{a} = \vec{a}_T + \vec{a}_N = a_T\hat{u}_T + a_N\hat{u}_N \qquad (3.22)$$

where

$$a_T = \frac{dv}{dt}, \quad a_N = \frac{v^2}{\rho} \qquad (3.23)$$

($v = \pm|\vec{v}|$; $\rho = radius\ of\ curvature$). Combining (3.22) with Newton's law, we find an expression for the total (resultant) force $\vec{F}$ acting on a particle of mass m:

$$\vec{F} = m\vec{a} = \vec{F}_T + \vec{F}_N = F_T\hat{u}_T + F_N\hat{u}_N \qquad (3.24)$$

where

Fig. 3.4 Tangential and normal components of acceleration and force

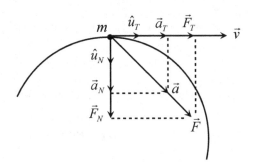

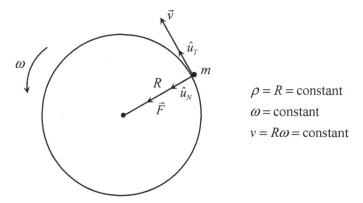

Fig. 3.5 A particle executing uniform circular motion

$$F_T = ma_T = m\frac{dv}{dt}, \quad F_N = ma_N = m\frac{v^2}{\rho} \qquad (3.25)$$

The *tangential* component $\vec{F}_T$ is responsible for the change of *speed*, while the *normal* (or *centripetal*) component $\vec{F}_N$ is responsible for the change of *direction* of motion.

If the total force $\vec{F}$ is *normal* to the trajectory (that is, perpendicular to the velocity) of the particle, then $F_T = 0$ and, according to the first of relations (3.25), the *speed* of the particle is constant in time (although the direction of the velocity does change). In other words, the particle executes *uniform curvilinear motion*. A typical example is *uniform circular motion*, in which the total force $\vec{F}$ is purely centripetal and always passes through the center of the circular path (see Fig. 3.5). By (3.24) and (3.25), and by using (2.34), we have:

$$\vec{F} = m\frac{v^2}{R}\hat{u}_N = mR\omega^2\hat{u}_N \qquad (3.26)$$

Exercise: A particle moves with constant speed on a plane curve, under the action of a total force of constant magnitude. Show that the particle performs uniform circular motion. (*Hint:* What can you conclude regarding the radius of curvature?)

3.7 Angular Momentum and Torque

Consider a particle of mass m, moving along some curve in space (see Fig. 3.6). The instantaneous position of the particle is determined by the position vector $\vec{r}$ relative to the fixed origin O of an inertial reference frame (we recall that it is in such frames

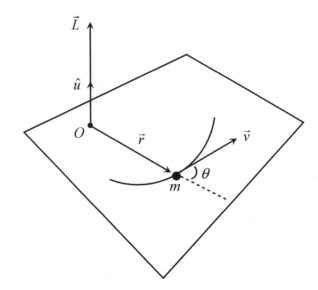

Fig. 3.6 The angular momentum vector is perpendicular to the momentary plane defined by the position vector and the velocity

only that Newton's laws are valid). Let $\vec{v}$ be the velocity of the particle at some point of the trajectory. The momentum of the particle at this point is $\vec{p} = m\vec{v}$.

The *angular momentum* of the particle *relative to point O* is defined as the cross product

$$\boxed{\vec{L} = \vec{r} \times \vec{p} = m(\vec{r} \times \vec{v})} \tag{3.27}$$

Note that, in contrast to the momentum, the angular momentum $\vec{L}$ is not an absolute quantity since its value depends on the choice of the reference point O.

The vector $\vec{L}$ is perpendicular to the instantaneous plane defined by $\vec{r}$ and $\vec{v}$, its direction being determined by the right-hand rule; that is, if we rotate the fingers of our right hand in the direction of instantaneous rotation of m about O, our extended thumb points in the direction of $\vec{L}$ (cf. Section 1.5). If θ is the angle between $\vec{r}$ and $\vec{v}$ (where $0 \leq \theta \leq \pi$) and if r and v are the respective magnitudes of these two vectors, the magnitude of the angular momentum is given by

$$|\vec{L}| = mrv \sin \theta \tag{3.28}$$

We define a unit vector $\hat{u}$ normal to the plane of $\vec{r}$ and $\vec{v}$ the direction of which vector ("up" or "down") is chosen arbitrarily. We can then write:

$$\vec{L} = \pm|\vec{L}|\hat{u} \equiv L\hat{u} \tag{3.29}$$

where L is the *algebraic value* of $\vec{L}$ with respect to $\hat{u}$. Note that, on the basis of the right-hand rule, the unit vector $\hat{u}$ defines a positive direction of momentary rotation

Fig. 3.7 The perpendicular distance of the reference point O from the axis of the velocity is $l = r(\sin\theta)$

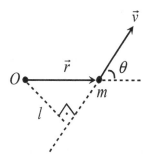

about O. Thus, in Fig. 3.6 the *counterclockwise* motion about O is in the positive direction since we chose the direction of $\hat{u}$ *upward* (if we had chosen the downward direction for $\hat{u}$, then it would be the clockwise motion about O the one in the positive direction). In Fig. 3.6 the particle m moves in the positive direction, so that $L > 0$.

An alternative expression for the magnitude of angular momentum is found with the aid of Fig. 3.7. We notice that $r \sin\theta = l$, where l is the perpendicular distance of O from the axis of $\vec{v}$. Equation (3.28) is thus written:

$$|\vec{L}| = mvl \tag{3.30}$$

(Note that the vector $\vec{L}$ is normal to the page and directed *outward*, i.e., toward the reader. What would be the direction of $\vec{L}$ if we reversed the direction of $\vec{v}$?)

The components of angular momentum are found with the aid of (1.23). If (x, y, z) and (p_x, p_y, p_z) are the components of $\vec{r}$ and $\vec{p}$, respectively, then

$$\vec{L} = \vec{r} \times \vec{p} = \begin{vmatrix} \hat{u}_x & \hat{u}_y & \hat{u}_z \\ x & y & z \\ p_x & p_y & p_z \end{vmatrix} = L_x\hat{u}_x + L_y\hat{u}_y + L_z\hat{u}_z$$

where

$$L_x = yp_z - zp_y, \quad L_y = zp_x - xp_z, \quad L_z = xp_y - yp_x \tag{3.31}$$

In particular, if the motion takes place on the xy-plane then $z = 0$ and $p_z = 0$, so that $L_x = L_y = 0$ and $\vec{L}$ is parallel to the z-axis.

Consider now the case where the particle executes *circular motion* of radius R about O (Fig. 3.8). We notice that $\theta = \pi/2$ and $r = l = R$. The angular momentum $\vec{L}$ of m with respect to O is a vector normal to the plane of the circle and directed in accordance with the direction of motion. By (3.28) or (3.30), and by the relation $v = R\omega$, we have:

$$\left|\vec{L}\right| = mRv = mR^2\omega \tag{3.32}$$

Fig. 3.8 Angular
momentum of a particle
executing circular motion

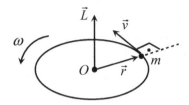

Returning to general curvilinear motion, we let $\vec{F}$ be a force acting on m at a point of the trajectory with radius vector $\vec{r}$ (see Fig. 3.9). The *torque* of $\vec{F}$ *relative* to the origin O of our inertial frame is defined as the cross product

$$\boxed{\vec{T} = \vec{r} \times \vec{F}} \tag{3.33}$$

The vector $\vec{T}$ is normal to the plane of $\vec{r}$ and $\vec{F}$ its direction being determined by the right-hand rule; that is, if we rotate the fingers of our right hand in the direction that $\vec{F}$ would *tend* to rotate m about O if m were at rest, then our extended thumb will point in the direction of $\vec{T}$. If θ is the angle between $\vec{r}$ and $\vec{F}$ (where $0 \leq \theta \leq \pi$) and if r and F are the respective magnitudes of these two vectors, the magnitude of the torque is

$$|\vec{T}| = rF \sin \theta \tag{3.34}$$

Also, if $\hat{u}$ is a unit vector normal to the plane of $\vec{r}$ and $\vec{F}$ (the direction of which vector is chosen arbitrarily), we write:

$$\vec{T} = \pm|\vec{T}|\hat{u} \equiv T\hat{u} \tag{3.35}$$

where T is the algebraic value of $\vec{T}$ with respect to $\hat{u}$.

Fig. 3.9 Torque is normal to
the plane defined by the
position vector and the force

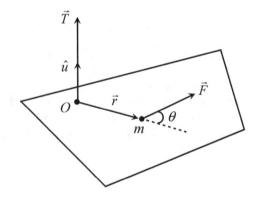

Fig. 3.10 The perpendicular distance of the reference point O from the axis of the force is $l = r(\sin\theta)$

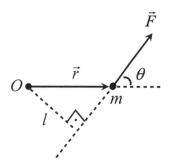

An alternative expression for the magnitude of torque is found with the aid of Fig. 3.10. Given that $r \sin \theta = l$, where l is the perpendicular distance of O from the axis of $\vec{F}$, we rewrite (3.34) as

$$|\vec{T}| = Fl \tag{3.36}$$

Finally, the components of $\vec{T}$ are (show this)

$$T_x = yF_z - zF_y, \quad T_y = zF_x - xF_z, \quad T_z = xF_y - yF_x \tag{3.37}$$

Between angular momentum and total torque there exists a relation analogous to the relation $\vec{F} = d\vec{p}/dt$ between linear momentum and total force. Differentiating the angular momentum (3.27) with respect to t, we have:

$$\frac{d\vec{L}}{dt} = \frac{d}{dt}(\vec{r} \times \vec{p}) = \frac{d\vec{r}}{dt} \times \vec{p} + \vec{r} \times \frac{d\vec{p}}{dt}$$

(note carefully that, upon differentiation, the order in which $\vec{r}$ and $\vec{p}$ appear must be preserved, since the cross product is not commutative). But,

$$\frac{d\vec{r}}{dt} \times \vec{p} = \vec{v} \times \vec{p} = \vec{v} \times (m\vec{v}) = m(\vec{v} \times \vec{v}) = 0.$$

Furthermore, by Newton's law (3.2) and by the definition (3.33) of torque,

$$\vec{r} \times \frac{d\vec{p}}{dt} = \vec{r} \times \vec{F} = \vec{T}$$

where $\vec{F}$ is assumed here to be the *resultant* force on m. Thus, finally,

$$\boxed{\vec{T} = \frac{d\vec{L}}{dt}} \tag{3.38}$$

Exercise: Show that the torque $\vec{T}$ of the resultant force on m is equal to the vector sum of the torques of all forces acting on m (cf. Appendix A).

Note that, in Eq. (3.38), $\vec{L}$ and $\vec{T}$ must be taken *with respect to the same point O*, which is the origin of our inertial frame of reference. Note also that (3.38) is an immediate consequence of Newton's law (3.2); it does not represent a new fundamental principle of Mechanics.

In the case where $\vec{T} = 0$, (3.38) yields:

$$\frac{d\vec{L}}{dt} = 0 \Rightarrow \vec{L} = \text{constant.}$$

This leads us to the *principle of conservation of angular momentum*:

When the torque of the total force on a particle, relative to some point, is zero, the angular momentum of the particle relative to this point is constant in time.

Even if such a point exists, however, there may be other points relative to which neither the torque of the total force vanishes, nor the angular momentum is constant!

Exercise: An electric charge Q is permanently located at the origin O of our coordinate system, while another charge q may move freely in space. Ignoring the weight of q, as well as the air resistance, show that the angular momentum of q relative to O is constant. Would the same be true for the angular momentum of q relative to a different point O' ? (Remember that the location of Q is fixed at O!)

3.8   Central Forces

Consider a particle of mass m, moving on a curved path under the action of a total force $\vec{F}$. The instantaneous position of m is determined by the position vector $\vec{r}$ with respect to the origin O of an inertial reference frame. In general, the force $\vec{F}$ varies in space (and, in particular, along the path of m). This force is thus a function of $\vec{r}$. We say that the particle m is moving in a *force field* $\vec{F} = \vec{F}(\vec{r})$.

Imagine now that we are able to choose a reference point O (see Fig. 3.11) such that

Fig. 3.11 A particle subject to a central force

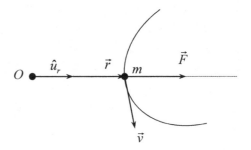

1. the line of action of $\vec{F}$ always passes through O, regardless of the position of the particle m in space;
2. the magnitude of $\vec{F}$ depends only on the distance $r = |\vec{r}|$ of m from O.

By defining the unit vector $\hat{u}_r$ in the direction of $\vec{r}$,

$$\hat{u}_r = \frac{\vec{r}}{|\vec{r}|} = \frac{\vec{r}}{r},$$

we can express both these conditions mathematically as follows:

$$\vec{F} = F(r)\hat{u}_r = \frac{F(r)}{r}\vec{r} \tag{3.39}$$

where $F(r) = \pm|\vec{F}|$ is an algebraic value, the sign of which depends on the relative orientation of $\vec{F}$ with respect to $\vec{r}$. A force (more correctly, a force field) of the form (3.39) is called a *central force* with center at O.

The motion of a particle m under the action of a central force has the following characteristics:

1. *The angular momentum $\vec{L}$ of the particle, with respect to the center O of the force, is constant during the motion of m.*
2. *The motion takes place on a constant plane.*

Indeed: The torque of $\vec{F}$ with respect to O is

$$\vec{T} = \vec{r} \times \vec{F} = \vec{r} \times \frac{F(r)}{r}\vec{r} = \frac{F(r)}{r}(\vec{r} \times \vec{r}) = 0.$$

Then, according to (3.38), the angular momentum $\vec{L}$ with respect to O is constant. Furthermore, the vector $\vec{L}$ is normal to the plane defined by $\vec{r}$ and $\vec{v}$. The constancy of $\vec{L}$, then, means that the aforementioned plane is constant as well. We thus conclude that the motion takes place on a constant plane.

A familiar example of a central force is the Coulomb force $\vec{F}$ experienced by an electric charge q inside the electrostatic field created by another charge Q located at some fixed point O (Fig. 3.12). According to Coulomb's law [2],

$$\vec{F} = \frac{1}{4\pi\varepsilon_0}\frac{Qq}{r^2}\hat{u}_r \equiv F(r)\hat{u}_r \tag{3.40}$$

Fig. 3.12 Coulomb force on an electric charge q due to a charge Q located at O

As mentioned previously (cf. Exercise at the end of Sect. 3.7) the angular momentum of q with respect to O is constant during the motion of this charge in the field of Q.

References

1. C.J. Papachristou, Foundations of newtonian dynamics: an axiomatic approach for the thinking student. Nausivios Chora **4**, 153 (2012). http://nausivios.snd.edu.gr/docs/partC2012.pdf; new version: https://arxiv.org/abs/1205.2326
2. C.J. Papachristou, *Introduction to Electromagnetic Theory and the Physics of Conducting Solids* (Springer, 2020)

Chapter 4
Work and Energy

Abstract The concepts of work and kinetic energy are defined and the work-energy theorem is proven. Conservative forces are introduced and conservation of mechanical energy is discussed. Examples of conservative forces are given.

4.1 Introduction

In principle, by Newton's second law we can predict the motion of a particle at any moment t, provided that we are given (a) the position and the velocity of the particle at time $t = 0$, and (b) the force field within which the motion of the particle takes place. In reality, however, the solution of the problem is not always easy, since Newton's law is not just a simple vector equation (don't be deceived by its innocent-looking form $\vec{F} = m\vec{a}$!) but is a *system of differential equations*:

$$m\frac{d\vec{v}}{dt} = \vec{F}(\vec{r}), \quad \frac{d\vec{r}}{dt} = \vec{v}.$$

By integrating this system for given *initial conditions* ($\vec{r} = \vec{r}_0$ and $\vec{v} = \vec{v}_0$ for $t = 0$) we determine the position and the velocity of the particle at all $t > 0$.

The difficulty in solving the problem directly (with the exception of a few simple cases) compels one to seek mathematical devices, most important of which—though applicable under specific conditions—is the *principle of conservation of mechanical energy*. This principle is an immediate consequence of Newton's second law; that is, it does not represent a new, independent postulate of Mechanics.

An even more general principle—also a consequence of Newton's law—is the *work-energy theorem*. In addition to its theoretical value, this theorem is particularly useful in cases where frictional forces are present and, therefore, conservation of mechanical energy cannot be applied.

C. J. Papachristou, *Introduction to Mechanics of Particles and Systems*,
https://doi.org/10.1007/978-3-030-54271-9_4

4.2 Work of a Force

Consider a particle of mass m moving in a *force field* $\vec{F}(\vec{r})$, where $\vec{r}$ is the position vector of m relative to the origin of an inertial reference frame (Fig. 4.1). Other forces, not belonging to the above field, may also act on the particle. We call $\vec{dr}$ the infinitesimal vector representing an elementary displacement of m along its trajectory, during an infinitesimal time interval dt. As dt approaches zero, $d\vec{r}$ tends to become tangent to the trajectory; that is, it tends to acquire the direction of the velocity $\vec{v}$ of the particle.

We define the *elementary work* of the force $\vec{F}$, for the displacement $d\vec{r}$ of m, as the dot product

$$dW = \vec{F} \cdot d\vec{r} = Fds \cos\theta \tag{4.1}$$

where $F = \left|\vec{F}\right|$ and $ds = |d\vec{r}|$, with the understanding that, within an infinitesimal time interval, ds may approximately be considered equal to the length of an infinitesimal section of the curved path of m. We say that $\vec{F}$ *produces work* dW in the time interval dt. Depending on the sign of $\cos\theta$, this work is positive if $0 \leq \theta < \pi/2$ and negative if $\pi/2 < \theta \leq \pi$ Note, in particular, that

a force normal to the velocity does not produce work,

since, in this case, $\theta = \pi/2$ and $\cos\theta = 0$, so that $dW = 0$.

We recall that in *uniform* curvilinear motion the speed of a particle m is constant and the *total* force $\vec{F}$ on m is purely centripetal, i.e., normal to the velocity. We thus conclude that

in uniform motion the resultant force $\vec{F}$ on a particle m does not produce work during the motion of m.

We now consider a finite part AB of the trajectory of m. We divide the curved path AB into a very large number of infinitesimal displacements $d\vec{r}_1, d\vec{r}_2, \ldots$, along which the values of the force field $\vec{F}(\vec{r})$ that acts on the particle are $\vec{F}_1, \vec{F}_2, \ldots$, as shown in Fig. 4.2. The elementary works produced along these displacements are

$$dW_1 = \vec{F}_1 \cdot d\vec{r}_1, \quad dW_2 = \vec{F}_2 \cdot d\vec{r}_2, \ldots.$$

Fig. 4.1 Trajectory of a particle moving in a force field

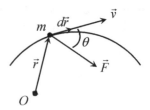

Fig. 4.2 The path *AB* in the
force field is divided into a
very large number of
infinitesimal displacements

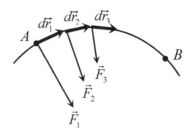

The *total work* of $\vec{F}(\vec{r})$ from *A* to *B* is

$$W = dW_1 + dW_2 + \cdots = \vec{F}_1 \cdot d\vec{r}_1 + \vec{F}_2 \cdot d\vec{r}_2 + \cdots = \sum_i \vec{F}_i \cdot d\vec{r}_i.$$

Since the displacements are infinitesimal, we may replace the sum with an integral:

$$\boxed{W = \int_A^B \vec{F}(\vec{r}) \cdot d\vec{r}} \tag{4.2}$$

Integrals of the form (4.2) are called *line integrals* since their value depends, in general, on the curve connecting two given points *A* and *B*. Hence, *the work of a force on a particle traveling from A to B is dependent upon the path followed by this particle from A to B* (an infinite number of such paths exist for given points *A* and *B*). In other words, different trajectories from *A* to *B* correspond to different values of the work of $\vec{F}$. The case of *conservative forces*, to be introduced later, constitutes an exception to this rule.

Let us see a simple example. In Fig. 4.3 the body moves on a straight line from *A* to *B* under the action of a *constant* (in magnitude and direction) force $\vec{F}$. Additional forces, not drawn in the figure, act on the body. Such forces are the weight of the body as well as the normal reaction and the kinetic friction from the plane where the motion takes place. We are interested here in the work of $\vec{F}$ alone, from *A* to *B*. Taking into account that both the magnitude *F* of $\vec{F}$, and the angle θ, are constant quantities, we have:

$$W = \int_A^B \vec{F} \cdot d\vec{r} = \int_A^B F \cos\theta\, ds = F\cos\theta \int_A^B ds \Rightarrow$$
$$W = Fs\cos\theta \tag{4.3}$$

Fig. 4.3 Rectilinear motion
of a body under the action of
a constant force (among
other forces, not drawn)

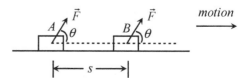

Fig. 4.4 A particle subject
to the action of several forces

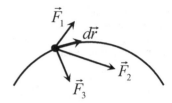

where s is the distance AB traveled by the body. In particular, $W = Fs$ if $\vec{F}$ is in the
direction of motion ($\theta = 0$), while $W = -Fs$ if $\vec{F}$ is opposite to the direction of motion
($\theta = \pi$).

When a particle is subject to the simultaneous action of several forces $\vec{F}_1, \vec{F}_2, \ldots$,
as in Fig. 4.4, *the work of the resultant force $\vec{F}$ equals the sum of the works of the
component forces.*

Proof The works of $\vec{F}_1, \vec{F}_2, \ldots$, for a displacement $d\vec{r}$ of the particle, are

$$dW_1 = \vec{F}_1 \cdot d\vec{r}, \quad dW_2 = \vec{F}_2 \cdot d\vec{r}, \ldots.$$

The work of the resultant force $\vec{F} = \vec{F}_1 + \vec{F}_2 + \cdots$, for the same displacement, is

$$dW = \vec{F} \cdot d\vec{r} = \left(\vec{F}_1 + \vec{F}_2 + \cdots \right) \cdot d\vec{r} = \vec{F}_1 \cdot d\vec{r} + \vec{F}_2 \cdot d\vec{r} + \cdots = dW_1 + dW_2 + \cdots ; \text{q.e.d.}$$

Now, let dt be the infinitesimal time interval within which the displacement $d\vec{r}$ of
a particle takes place, and let $\vec{F}$ be the force acting on this particle in this interval
(for an infinitesimal dt, $\vec{F}$ can be assumed to be constant). The elementary work of
$\vec{F}$ in the interval dt is $dW = \vec{F} \cdot d\vec{r}$. The *work per unit time* produced by $\vec{F}$ is equal
to

$$P = \frac{dW}{dt} \tag{4.4}$$

and is called the *power* supplied by the agent that exerts the force $\vec{F}$. We have:

$$P = \frac{\vec{F} \cdot d\vec{r}}{dt} = \vec{F} \cdot \frac{d\vec{r}}{dt} \Rightarrow$$

$$\boxed{P = \vec{F} \cdot \vec{v}} \tag{4.5}$$

where $\vec{v}$ is the instantaneous velocity of the particle. The work produced by $\vec{F}$ in the
time interval between t_1 and t_2 is

$$W = \int_{t_1}^{t_2} P dt = \int_{t_1}^{t_2} (\vec{F} \cdot \vec{v}) dt \tag{4.6}$$

In S.I. units the unit of work is 1 *Joule* $(1 \ J) = 1 \ N \ m = 1 \ kg \ m^2 \ s^{-2}$, while the unit of power is 1 W $(1 \ W) = 1 \ J \ s^{-1} = 1 \ kg \ m^2 \ s^{-3}$. The units $1 \ kW = 10^3 \ W$ and $1 \ MW$ $= 10^6 \ W$ are also used in applications.

4.3 Kinetic Energy and the Work-Energy Theorem

In the previous section we stated that a force normal to the velocity does not produce work. On the other hand, a *resultant* force normal to the velocity cannot produce a change of *speed*. One is thus led to expect that there must be some relation between the work of the total force and the change of speed, so that the vanishing of work implies the constancy of speed, and vice versa.

We define the *kinetic energy* of a particle of mass m moving at a speed v:

$$E_k = \frac{1}{2}mv^2 \tag{4.7}$$

If $p = mv$ is the magnitude of the particle's momentum, relation (4.7) can also be written in the equivalent form,

$$E_k = \frac{p^2}{2m} \tag{4.8}$$

Now, let $\vec{F}$ be the *resultant* force on a particle m. By Newton's law,

$$\vec{F} = m\vec{a} = m\frac{d\vec{v}}{dt}.$$

The elementary work of $\vec{F}$ for a displacement $d\vec{r}$ of the particle is

$$dW = \vec{F} \cdot d\vec{r} = m\frac{d\vec{v}}{dt} \cdot d\vec{r} = m\frac{d\vec{r}}{dt} \cdot d\vec{v} = m\vec{v} \cdot d\vec{v}.$$

But,

$$\vec{v} \cdot d\vec{v} = \frac{1}{2}d(\vec{v} \cdot \vec{v}) = \frac{1}{2}d(v^2) = \frac{1}{2}\frac{d(v^2)}{dv}dv = \frac{1}{2}(2v)dv = vdv$$

where $v = |\vec{v}|$ is the speed of the particle and where we have used (1.17). Thus, dW $= mvdv$. The work of $\vec{F}$ for a finite displacement of m from point A to point B along its trajectory is found by integrating:

$$W = \int_A^B dW = m \int_A^B v dv = m \left[\frac{v^2}{2} \right]_A^B \Rightarrow$$

$$\boxed{W = \frac{1}{2}mv_B^2 - \frac{1}{2}mv_A^2} \tag{4.9}$$

Combining this with (4.7), we have:

$$\boxed{W = E_{k,B} - E_{k,A} \equiv \Delta E_k} \tag{4.10}$$

Relations (4.9) and (4.10) constitute the mathematical expression of the *work-energy theorem*, which states the following:

The work of the total force on a particle (equal to the sum of works of all forces acting on the particle), in a displacement of the particle from one point of its trajectory to another, is equal to the change of the kinetic energy of the particle in this displacement.

In particular, if the work of the resultant force on the particle is zero, the kinetic energy of the particle is constant and, by (4.7), the same is true for the particle's speed. This is what happens in uniform rectilinear motion, where the total force is zero, as well as in uniform curvilinear motion, where the total force is perpendicular to the velocity. In both cases the total work vanishes and the speed of the particle is constant.

Note that the work-energy theorem is an immediate consequence of Newton's law; it does not express a new, independent principle of Mechanics. According to the definition (4.7), the unit of kinetic energy is $1 \ kg \ m^2 \ s^{-2} = (1 \ kg \ m \ s^{-2}) \ (1 \ m) = 1 \ N \ m = 1 \ J$. That is, E_k is measured in units of work, as was to be expected in view of (4.10).

4.4 Potential Energy and Conservative Forces

Consider a particle of mass m subject to a force $\vec{F}$ in some region of space. In general, $\vec{F}$ varies from point to point in that region, each point determined by its corresponding position vector $\vec{r}$ relative to the origin O of coordinates (x, y, z) of an inertial reference frame. We assume that the force $\vec{F}$ is dependent only on the position of m in the region of interest (which is not the case, e.g., with kinetic friction, the direction of which at any point depends on the direction of motion at that point). We write:

$$\vec{F} = \vec{F}(\vec{r}) = \vec{F}(x, y, z) \tag{4.11}$$

Strictly speaking, relation (4.11) represents a *force field* rather than just a single force. Nevertheless, we will continue to refer to "the force", for brevity.

The work of $\overrightarrow{F}$ when the particle m moves from point A to point B in space is

$$W = \int_A^B \overrightarrow{F}(\overrightarrow{r}) \cdot d\overrightarrow{r} \tag{4.12}$$

As stressed in Sect. 4.2, the value of the above integral is dependent not just on the limit points A and B but also on the *path* followed by m from A to B (there is an infinite number of paths connecting A with B).

There is, however, a special class of forces (more correctly, force fields) of the form (4.11), the work W of which depends *only* on the limit points A and B, regardless of the path joining them. Such forces are said to be *conservative*.

Definition A force of the form (4.11) is called *conservative* if a function $E_p(\overrightarrow{r}) = E_p(x, y, z)$ exists, such that the work of $\overrightarrow{F}$ from A to B is equal to the difference of the values of E_p at the points A and B. Explicitly,

$$W = \int_A^B \overrightarrow{F} \cdot d\overrightarrow{r} = E_p(\overrightarrow{r}_A) - E_p(\overrightarrow{r}_B) \equiv E_{p,A} - E_{p,B} \tag{4.13}$$

Given that the *change* of E_p from A to B is

$$\Delta E_p = E_{p,B} - E_{p,A} = \text{final value minus initial value,}$$

relation (4.13) is written briefly:

$$W = -\Delta E_p \tag{4.14}$$

The function $E_p(\overrightarrow{r})$ is called the *potential energy* of the particle m in the force field $\overrightarrow{F}(\overrightarrow{r})$ (we often say that the potential energy E_p is *associated with the conservative force* $\overrightarrow{F}$). As is obvious from (4.13), E_p is measured in units of work.

If several conservative forces act on m, each force is associated with a corresponding potential energy. As is easy to show, the potential energy associated with the *resultant* of a number of conservative forces is equal to the *algebraic sum* of the potential energies associated with the component forces. Hence, if $\overrightarrow{F}$ in (4.13) represents the total conservative force on m, then E_p represents the total potential energy of that particle in the force field $\overrightarrow{F}$.

If additional, *non-conservative* forces act on the particle m, these are *not* included in the force $\overrightarrow{F}$ appearing in relation (4.13), nor are they associated with some potential energy. In such a case, W in (4.13) represents the total work of the *conservative* forces only, *not* the work of the resultant of all forces acting on m.

We could have defined E_p differently in order for (4.14) to be written $W = +\Delta E_p$. This simply means putting $(-E_p)$ in place of E_p, i.e., defining E_p with the opposite sign. There is no particular physical consequence in doing this! The choice of the negative sign in (4.14) is purely a matter of convention in order for the total mechanical energy of m (see following section) to be expressed as a sum, rather than as a difference.

We also note the following:

1. Let $E_p(\vec{r})$ be the potential energy associated with the conservative force $\vec{F}(\vec{r})$. Then, the function

$$E'_p(\vec{r}) = E_p(\vec{r}) + C$$

where C is an arbitrary constant having dimensions of work, also is a potential energy for the same force $\overrightarrow{F}$. Indeed, if W is the work of $\overrightarrow{F}$, then

$$W = E_{p,A} - E_{p,B} = \left(E_{p,A} + C\right) - \left(E_{p,B} + C\right) = E'_{p,A} - E'_{p,B}.$$

We notice that the definition of potential energy allows for some degree of arbitrariness, since we can add any constant quantity to the function $E_p(\vec{r})$ without altering the physics of the situation (the force $\overrightarrow{F}$ is unaffected by this arbitrary addition to the potential energy). Because of this arbitrariness, we are free to choose *any point* (or *any plane*) *of reference* where the value of E_p is assumed to be zero.

2. By the definition (4.13) it follows that the work of a conservative force $\overrightarrow{F}$, when a particle travels from a point A to another point B, is *independent of the path connecting these points*. Hence, if C_1 and C_2 are two different paths joining A and B, and if W_1 and W_2 are the respective works of $\overrightarrow{F}$ along these paths, then

$$W_1 = W_2 = E_{p,A} - E_{p,B}.$$

But, wait a minute: the work-energy theorem (4.10) similarly tells us that $W = E_{k,B}-E_{k,A}$, which is valid even if the force $\vec{F}$ is *not* conservative! Careful, however: the difference ΔE_k *is dependent*, in general, upon the path joining A and B, whereas the difference ΔE_p does *not* depend on it. This is due to the fact that E_p is a given function of the position of the particle, which is generally not the case with E_k (remember that this latter quantity depends on speed, rather than on position). We stress that

the work-energy theorem, $W = \Delta E_k$ (where W is the work of the *resultant* force on the particle m) is of *general* validity, *regardless of the kind of forces acting on m*. On the contrary, the relation $W = -\Delta E_p$ concerns the work of *conservative* forces only and does *not necessarily* represent the total work done on m, the latter work containing possible additional contributions from non-conservative forces.

3. The work of a conservative force $\overrightarrow{F}$ along a *closed* path is *zero*. This can be shown as follows:

Consider a closed path C and two arbitrary points A, B on it (Fig. 4.5). Thus, the path C consists of two segments; namely, C_1 from A to B, and C_2 from B back to

Fig. 4.5 A closed path partitioned into two open paths: C_1 from A to B and C_2 from B back to A

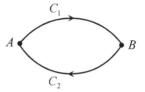

A. We write, symbolically, $C = C_1 + C_2$. Then, the work W along C is the sum of works W_1 and W_2 along C_1 and C_2, respectively:

$$W = W_1 + W_2 = \left(E_{p,A} - E_{p,B}\right) + \left(E_{P,B} - E_{p,A}\right) = 0.$$

In general, the work along a closed path is represented by a *closed line integral*. In particular, for a conservative force $\vec{F}$ we have:

$$W = \oint \vec{F} \cdot d\vec{r} = 0$$

where the little circle on the integral sign indicates that the path of integration is a closed curve.

We remark that the above-stated properties of conservative forces may in fact be regarded as definitions of the concept of a conservative force, equivalent to (4.13) (see, e.g., [1]).

4.5 Conservation of Mechanical Energy

Consider a particle of mass m moving from point A to point B in some region of space, under the action of a force (more correctly, a force field) $\vec{F}$ which is the resultant of all forces acting on m. According to the work-energy theorem, the work W of $\vec{F}$ equals the change of kinetic energy of m:

$$W = \Delta E_k = E_{k,B} - E_{k,A} \qquad (4.15)$$

We stress again that (4.15) gives the work of the *resultant* force, regardless of whether this force is conservative or not. Now, if it happens that the total force $\vec{F}$ is *conservative* (which means that *all* component forces on m are conservative) then the work of $\vec{F}$ can be expressed as

$$W = -\Delta E_p = E_{p,A} - E_{p,B} \qquad (4.16)$$

where E_p is the total potential energy of m. Equating the right-hand sides of (4.15) and (4.16), we have:

$$E_{k,A} + E_{p,A} = E_{k,B} + E_{p,B} \qquad (4.17)$$

Relation (4.17) is valid for any two points A and B between which the motion of m takes place. We thus conclude that the quantity

$$\boxed{E = E_k + E_p = \frac{1}{2}mv^2 + E_p(\vec{r})} \qquad (4.18)$$

which is called the *total mechanical energy* of m in the force field $\vec{F}$, is constant during the motion of m in that field.

The above conclusion leads us to the *principle of conservation of mechanical energy*:

> If all forces acting on a particle m are conservative, the total mechanical energy of m retains a constant value along the particle's trajectory.

We now see the logic behind the term "conservative force". For non-conservative forces (such as, e.g., kinetic friction; see Sect. 4.7) it is not possible to define a potential energy, thus also a total mechanical energy. Conservation of mechanical energy must therefore be reexamined when such forces are present.

So, how do we handle the case where a particle m is subject to the simultaneous action of *both* conservative *and* non-conservative forces (such as, e.g., gravity and air resistance, respectively, for a falling body)? Let $\vec{F}$ be the total conservative force on m, and let $\vec{F}'$ be the total non-conservative force on this particle. The resultant of all forces on m is

$$\vec{F}_{tot} = \vec{F} + \vec{F}'$$

and the work of this force from A to B is

$$W = \int_A^B \vec{F}_{tot} \cdot d\vec{r} = \int_A^B \vec{F} \cdot d\vec{r} + \int_A^B \vec{F}' \cdot d\vec{r} \qquad (4.19)$$

The first integral on the right is equal to the difference $(E_{p,A} - E_{p,B})$, where E_p is the potential energy associated with the conservative force $\vec{F}$, while the second integral represents the work W' of the non-conservative force $\vec{F}'$. Furthermore, by the work-energy theorem, the work W of the resultant force $\vec{F}_{tot}$ is equal to $(E_{k,B} - E_{k,A})$. Relation (4.19) is thus written as

$$E_{k,B} - E_{k,A} = \left(E_{p,A} - E_{p,B}\right) + W' \Rightarrow$$

$$W' = (E_{k,B} + E_{p,B}) - (E_{k,A} + E_{p,A}) \equiv \Delta(E_k + E_p) \qquad (4.20)$$

We therefore conclude that

if non-conservative forces are present, the sum $(E_k + E_p)$ is not constant, in general; specifically, this quantity changes by an amount equal to the work of the non-conservative forces.

An exception occurs when the non-conservative forces *do not produce work*, i.e., when $W' = 0$. This is the case when the total non-conservative force $\vec{F}'$ is normal to the velocity of m. It then follows from (4.20) that the sum $(E_k + E_p)$ is constant:

$$\Delta(E_k + E_p) = 0 \Leftrightarrow E_k + E_p = \quad \text{constant, when} \quad W' = 0.$$

We may thus express the principle of conservation of mechanical energy more generally, as follows:

If the non-conservative forces that act on a particle do not produce work, the total mechanical energy of the particle is constant.

For example, we apply conservation of mechanical energy in the study of the motion of a pendulum (see Problem 25) despite the fact that the bob of the pendulum is subject not only to the conservative force of gravity but also to the tension of the string. This tension does not produce work, however, since it is always perpendicular to the velocity of the bob (explain why).

Note: Eq. (4.20) may be interpreted more generally, as follows: Assume that the total force acting on a particle can be written as a sum of two forces, $\vec{F}$ and $\vec{F}'$, where $\vec{F}$ is *conservative* and is associated with some potential energy E_p, while $\vec{F}'$ may be of *any* kind, i.e., conservative, non-conservative, or with mixed conservative and non-conservative components. Then, according to (4.20), the work W' of $\vec{F}'$ equals the change of the sum $(E_k + E_p)$.

4.6 Examples of Conservative Forces

The following examples demonstrate a method for showing that a given force field is conservative. Specifically, we evaluate the work W of the force for an arbitrary path from A to B and then use Eq. (4.13) to read off the potential energy E_p, if it exists. The existence of E_p automatically proves the conservative property of the given force field.

1. *Force of gravity*

Near the surface of the Earth the acceleration $\vec{g}$ of gravity is practically constant over large regions of space. Hence, in accordance with the results of Sect. 2.5, the motion of a body under the sole action of gravity takes place in a constant plane—say, the

Fig. 4.6 Motion of a body
near the surface of the Earth,
under the sole action of
gravity

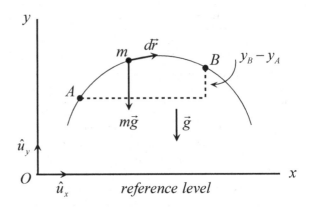

xy-plane—perpendicular to the surface of the Earth, where the *x*-axis is horizontal while the *y*-axis is vertical and, by arbitrary choice, directed upward (see Fig. 4.6). The *y*-coordinate specifies the instantaneous height at which a particle is located above an arbitrary horizontal reference level at which $y = 0$. With the chosen upward direction of the vertical axis, the particle is above or below the reference level if $y > 0$ or $y < 0$, respectively. In a 3-dimensional Cartesian coordinate system (x, y, z) the *z*-coordinate of the particle has the fixed value $z = 0$, where the *z*-axis (not shown in Fig. 4.6) is normal to the page and directed toward the reader.

The weight of a particle of mass *m* is a force perpendicular to the surface of the Earth (thus parallel to the *y*-axis) and directed downward:

$$\vec{F} = m\vec{g} = -mg\hat{u}_y \equiv (0, -mg, 0)$$

(the *x*- and *z*-components of $\vec{F}$ are zero). On the other hand, an elementary displacement in space is written, in general,

$$d\vec{r} = d(x\hat{u}_x + y\hat{u}_y + z\hat{u}_z) = (dx)\hat{u}_x + (dy)\hat{u}_y + (dz)\hat{u}_z \equiv (dx, dy, dz).$$

Here *z* is fixed, so that $dz = 0$. Thus, an elementary displacement of the particle on its trajectory is written:

$$d\vec{r} = (dx)\hat{u}_x + (dy)\hat{u}_y \equiv (dx, dy, 0).$$

Then, by using (1.19),

$$\vec{F} \cdot d\vec{r} = 0 \cdot dx + (-mg)dy + 0 \cdot 0 = -mgdy.$$

The work of $\vec{F}$ when *m* moves from *A* to *B* along its trajectory is

$$W = \int_A^B \vec{F} \cdot d\vec{r} = -\int_A^B mg\,dy = -mg \int_A^B dy = -mg(y_B - y_A) \Rightarrow$$

$$W = mgy_A - mgy_B \tag{4.21}$$

Is the force $\vec{F}$ conservative? In order for it to be, one should be able to find a function $E_p(\vec{r})$, the *potential energy of m in the gravitational field of the Earth*, such that

$$W = E_{p,A} - E_{p,B}.$$

By (4.21) we see that such a function indeed exists:

$$\boxed{E_p(y) = mgy} \tag{4.22}$$

More generally, $E_p = mgy + C$, where C is an arbitrary constant quantity. We can eliminate C by (arbitrarily) requiring that $E_p = 0$ at the reference level $y = 0$.

Exercise: Show that, if we choose the *downward* direction for the y-axis, then relation (4.22) must be rewritten as $E_p(y) = -mgy$.

According to the principle of conservation of mechanical energy, the total mechanical energy of m remains fixed as the particle moves under the action of gravity (if we ignore non-conservative forces such as air resistance):

$$E = E_k + E_p = \frac{1}{2}mv^2 + mgy = const. \tag{4.23}$$

Equivalently, for any two points A and B,

$$E_A = E_B \Leftrightarrow \frac{1}{2}mv_A^2 + mgy_A = \frac{1}{2}mv_B^2 + mgy_B \tag{4.24}$$

Exercise: By using (4.24) show that, in a free fall from a height h, a body acquires a speed equal to

$$v = \sqrt{2gh}.$$

Will this result be valid if air resistance is taken into account?

2. *Elastic force*

Consider a particle of mass m moving along the x-axis under the action of a force

$$\vec{F} = -kx\hat{u}_x \tag{4.25}$$

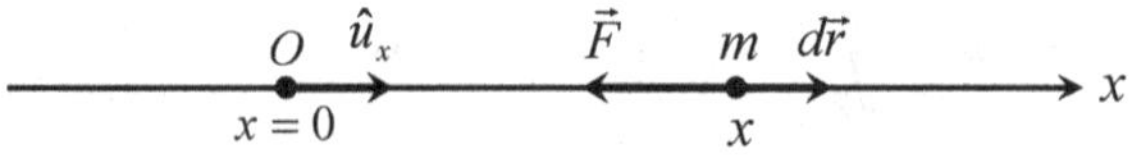

Fig. 4.7 A particle subject
to an elastic force

where k is a positive constant (Fig. 4.7). A force of the type (4.25) is called *elastic force*; its physical significance will be studied in Chap. 5.

The elementary displacement of m on the x-axis is written:

$$d\vec{r} = (dx)\hat{u}_x.$$

Thus, by applying (1.19) for two vectors having zero y- and z-components,

$$\vec{F} \cdot d\vec{r} = -kxdx.$$

The work of $\vec{F}$ for a displacement of m from point A to point B on the x-axis is

$$W = \int_A^B \vec{F} \cdot d\vec{r} = -\int_A^B kxdx = \frac{1}{2}kx_A^2 - \frac{1}{2}kx_B^2.$$

Now, in order for $\vec{F}$ to be conservative, one should be able to find a potential energy function E_p such that

$$W = E_{p,A} - E_{p,B}.$$

Indeed:

$$\boxed{E_p(x) = \frac{1}{2}kx^2} \tag{4.26}$$

where we have arbitrarily demanded that $E_p = 0$ at $x = 0$. By conservation of mechanical energy,

$$E = E_k + E_p = \frac{1}{2}mv^2 + \frac{1}{2}kx^2 = const. \tag{4.27}$$

3. *Coulomb force*

The force on an electrically charged particle of charge q, inside the electrostatic field created by another charge Q (Fig. 4.8), is given by Eq. (3.40):

$$\vec{F} = k\frac{Qq}{r^2}\hat{u}_r \equiv F(r)\hat{u}_r \tag{4.28}$$

where $k = 1/4\pi \varepsilon_0$ in S.I. units. Given that $\hat{u}_r = \vec{r}/r$ (where $r = |\vec{r}|$), (4.28) is written:

Fig. 4.8 Motion of an electric charge q in the Coulomb field produced by another charge Q

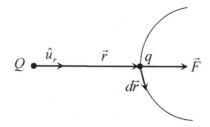

$$\vec{F} = \frac{F(r)}{r}\vec{r}.$$

If $d\vec{r}$ is an elementary displacement of q on its trajectory,

$$\vec{F} \cdot d\vec{r} = \frac{F(r)}{r}\vec{r} \cdot d\vec{r}.$$

But,

$$\vec{r} \cdot d\vec{r} = \frac{1}{2}[(d\vec{r}) \cdot \vec{r} + \vec{r} \cdot d\vec{r}] = \frac{1}{2}d(\vec{r} \cdot \vec{r}) = \frac{1}{2}d\left(r^2\right) = \frac{1}{2}\frac{d\left(r^2\right)}{dr}dr = \frac{1}{2}(2r)dr = rdr.$$

Hence,

$$\vec{F} \cdot d\vec{r} = \frac{F(r)}{r}rdr = F(r)dr.$$

The work of $\vec{F}$ for a displacement of q from point A to point B on its trajectory is

$$W = \int_A^B \vec{F} \cdot d\vec{r} = \int_A^B F(r)dr = kQq\int_A^B \frac{dr}{r^2} = kQq\left(\frac{1}{r_A} - \frac{1}{r_B}\right) \Rightarrow$$

$$W = k\frac{Qq}{r_A} - k\frac{Qq}{r_B} \tag{4.29}$$

The force $\vec{F}$ is conservative if there exists a *potential energy* E_p of q in the *Coulomb field of Q*, such that $W = E_{p,A} - E_{p,B}$. The function E_p is easy to read off from (4.29):

$$\boxed{E_p(r) = k\frac{Qq}{r}} \tag{4.30}$$

where we have arbitrarily assumed that $E_p = 0$ at an infinite distance from Q, i.e., for $r = \infty$. We conclude that the electrostatic Coulomb force is a conservative force.

Fig. 4.9 Kinetic friction is always directed opposite to the velocity of the moving body

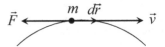

Note: Due to the symmetry of (4.30), this relation equally expresses the potential energy of Q in the Coulomb field of q. For this reason one says that (4.30) represents the potential energy of the *system* of charges Q and q (see [2], Chap. 5).

4.7   Kinetic Friction as a Non-Conservative Force

Kinetic friction (we here denote it $\vec{F}$) is always opposite to the velocity $\vec{v}$ of a particle, thus opposite to the elementary displacement $d\vec{r}$ of the particle on its trajectory (see Fig. 4.9). This means that the elementary work of $\vec{F}$ for the displacement $d\vec{r}$ is always negative:

$$dW = \vec{F} \cdot d\vec{r} < 0.$$

Hence, the work of $\vec{F}$ along *any* path is negative. In particular, for a path represented by a closed curve we have:

$$\oint \vec{F} \cdot d\vec{r} < 0.$$

That is, the closed line integral of $\vec{F}$ is different from zero. According to what was said in Sect. 4.4, this result indicates that *kinetic friction cannot be a conservative force.*

It is meaningless to ask whether *static* friction is or is not conservative, given that this force does not produce work (why?). In Chap. 7 we will see that we may use conservation of mechanical energy for rolling bodies in spite of the presence of static friction.

References

1. C.J. Papachristou, *Aspects of Integrability of Differential Systems and Fields: A Mathematical Primer for Physicists* (Springer, 2019)
2. C.J. Papachristou, *Introduction to Electromagnetic Theory and the Physics of Conducting Solids* (Springer, 2020)

Chapter 5
Oscillations

Abstract Simple harmonic motion (SHM) is studied and shown to be intimately related to uniform circular motion. Oscillations of a mass-spring system, as well as of a pendulum, are studied. The differential equation of SHM is derived.

5.1 Simple Harmonic Motion (SHM)

In this chapter we will study a special type of rectilinear motion called *simple harmonic motion* (SHM) or *harmonic oscillation*. It is a kind of *periodic motion*, in the sense that it consists of a continuous repetition of a certain prototype motion or "cycle". This motion may describe a physical *vibration*, such as that of a mass-spring system or, to a good approximation, of a pendulum.

The motion is confined to a section of the x-axis limited by the points $x = -A$ and $x = +A$ (Fig. 5.1). The *displacement* x of the moving body relative to the origin O is given as a function of time t by an expression of the form

$$x = A \cos(\omega t + \alpha) \tag{5.1}$$

where A and ω are positive constants while α is a constant that may assume any real value. The constants A and ω represent the *amplitude* and the *angular frequency*, respectively, of the SHM, while the angle $\omega t + \alpha$ (in *rad*) is called the *phase*. The angular frequency ω has dimensions of inverse time in order for ωt to be dimensionless.

Note that we could have described the same motion by using a sine function instead of a cosine one:

$$x = A \sin(\omega t + \beta) \tag{5.2}$$

This, however, reduces to the cosine form (5.1) by setting $\beta = \alpha + \pi/2$.

An effective way to visualize the kind of motion described by (5.1) is the following (see Fig. 5.2). Imagine a particle executing uniform circular motion on the xy-plane,

C. J. Papachristou, *Introduction to Mechanics of Particles and Systems*,
https://doi.org/10.1007/978-3-030-54271-9_5

69

Fig. 5.1 The oscillatory
motion of the body takes
place between $x = -A$ and x
$= +A$

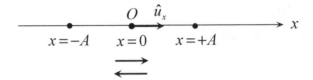

Fig. 5.2 As the point P
performs uniform circular
motion, its projection
x executes SHM between $-A$
and $+A$

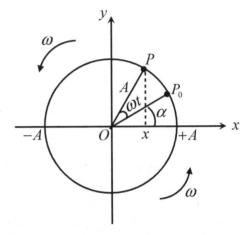

with constant angular velocity ω. We call A the radius of the circular trajectory and
we assume that the motion is *counterclockwise*. The center O of the circle coincides
with the origin of our coordinate system; thus, this circle intersects the x-axis at the
points $x = -A$ and $x = +A$. At the moment $t = 0$ the particle is located at some given
point P_0 of the circle, while at the (arbitrary) moment t the particle passes from a
point P. Between the moments $t = 0$ and t the position vector of the particle sweeps
out an angle $P_0OP = \omega t$. Hence, the angle between the position vector and the x-axis
at time t is $\varphi(t) = \omega t + \alpha$ [see relation (2.39)].

As t increases indefinitely, the point P representing the instantaneous position of
the particle traces out the circle repeatedly with constant angular velocity ω. At the
same time, the *projection* of P onto the x-axis *oscillates* continuously along the axis
between the points $x = -A$ and $x = +A$. The value x of the projection of P at time t
is given by the relation

$$x(t) = A \cos \varphi(t) = A \cos(\omega t + \alpha) .$$

We thus see that, as the point P moves counterclockwise on the circle, with
constant angular velocity ω, the projection x of P executes SHM with angular
frequency ω and with amplitude A equal to the radius of the circle. Furthermore,
the instantaneous value of the angle $\varphi(t)$ gives the phase of the SHM. In particular,
the value $\varphi(0) = \alpha$ of the phase at $t = 0$ is called the *initial phase*.

The *period T* of a SHM is the time required in order for the oscillating body to pass from the same point x twice, *moving in the same direction*. Equivalently, T is the time required in order for the point P to describe a full circle, or, in order for the position vector of P to sweep out an angle 2π. Thus, the angular velocity of the circular motion, equal to the angular frequency of the SHM, is

$$\omega = \frac{2\pi}{T} \qquad (5.3)$$

[see (2.40)]. In the course of a period, the phase of the SHM increases by 2π. Indeed,

$$\varphi(t) = \omega t + \alpha \Rightarrow \varphi(t + T) = \omega(t + T) + \alpha = \varphi(t) + \omega T = \varphi(t) + 2\pi \ .$$

Thus, x returns to its initial value:

$$x(t + T) = A \cos \varphi(t + T) = A \cos[\varphi(t) + 2\pi] = A \cos \varphi(t) = x(t) \ .$$

For the particular value $\alpha = -\pi/2$ of the initial phase, relation (5.1) is graphically represented as in Fig. 5.3. Notice that the same graph describes the sine function $x = A \sin \omega t$, which is of the form (5.2) with $\beta = 0$.

Given that the oscillating body executes a complete oscillation within the time T of a period, how many oscillations does the body execute in the unit of time? Equivalently, how many complete revolutions does the point P make along the circle in the unit of time?

Assume that time is measured in some unit 1τ, where τ could mean seconds, minutes, days, months, years, etc. We think as follows:

Time T (measured in τ) corresponds to 1 oscillation or 1 revolution;
time 1τ corresponds to (say) N oscillations or N revolutions.

Then,

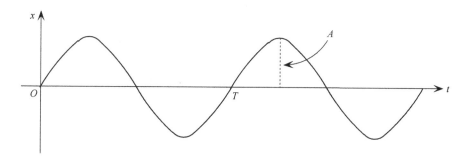

Fig. 5.3 Graph of the SHM (5.1) for $\alpha = -\pi/2$

$$\frac{T}{1\tau} = \frac{1}{N} \Rightarrow \frac{N}{1\tau} = \frac{1}{T}.$$

The quantity

$$f = \frac{1}{T} \tag{5.4}$$

is called the *frequency* of the SHM. In S.I. units, f is measured in s^{-1}. This unit is also called *hertz* (*Hz*) or *cycle per second*. By combining (5.3) and (5.4) we have:

$$\omega = \frac{2\pi}{T} = 2\pi f \tag{5.5}$$

The angular frequency ω is measured in $rad.s^{-1}$.

5.2 Force in SHM

Consider a body of mass m, performing SHM. Relation (5.1) gives the displacement x of m from the center of oscillation O, as a function of time t. The velocity and the acceleration of m are given by (2.1) and (2.3), respectively:

$$\vec{v} = v\hat{u}_x , \quad \vec{a} = a\hat{u}_x ,$$

where the algebraic values v and a of the two vectors are

$$v = \frac{dx}{dt} = -\omega A \sin(\omega t + \alpha) \tag{5.6}$$

and

$$a = \frac{dv}{dt} = -\omega^2 A \cos(\omega t + \alpha) = -\omega^2 x \tag{5.7}$$

By Newton's law, the *resultant* force on m is

$$\vec{F} = m\vec{a} = ma\hat{u}_x = -m\omega^2 x\hat{u}_x \equiv F\hat{u}_x .$$

The algebraic value F of the total force is

$$F = -kx \tag{5.8}$$

where we have put

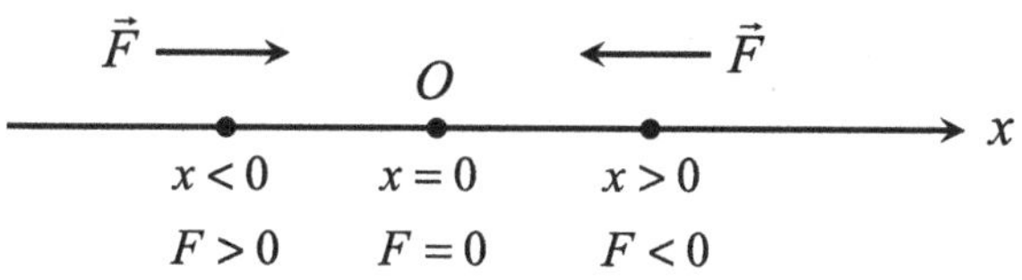

Fig. 5.4 Total force on a
body executing SHM

$$\vec{F} \longrightarrow \qquad O \qquad \longleftarrow \vec{F}$$

$$\xrightarrow{\hspace{10cm}} x$$

$$
\begin{array}{ccc}
x < 0 & x = 0 & x > 0 \\
F > 0 & F = 0 & F < 0
\end{array}
$$

$$k = m\omega^2 \Leftrightarrow \omega = \sqrt{\frac{k}{m}} \tag{5.9}$$

Relation (5.5) now yields:

$$f = \frac{\omega}{2\pi} = \frac{1}{2\pi}\sqrt{\frac{k}{m}}$$

$$\tag{5.10}$$

$$T = \frac{2\pi}{\omega} = 2\pi\sqrt{\frac{m}{k}}$$

From (5.8) we see that in SHM the total force is always *opposite* to the displacement from O and is directed toward O (see Fig. 5.4). The point O, where $x = 0$, is called the *equilibrium position* of SHM, since $F = 0$ there. This does *not necessarily* mean that the body is at rest at O but that the resultant force on it vanishes at that point. If, however, the body is initially at rest at the equilibrium position O, it will remain at rest there.

5.3 Energy Relations

With the aid of (5.6) we can find an expression for the kinetic energy of a body of mass m, executing SHM:

$$E_k = \frac{1}{2}mv^2 = \frac{1}{2}m\omega^2 A^2 \sin^2(\omega t + \alpha) = \frac{1}{2}m\omega^2\left[A^2 - A^2\cos^2(\omega t + \alpha)\right].$$

By using (5.1) and (5.9) we get:

$$E_k = \frac{1}{2}k\left(A^2 - x^2\right) \tag{5.11}$$

We notice that E_k obtains its maximum value at the equilibrium position ($x = 0$) and vanishes at the extreme points ($x = \pm A$) of the oscillation.

It follows from (5.8) that the *resultant* force $\vec{F}$ on the oscillating body is an *elastic force* of the form (4.25). As we showed in Sect. 4.6 [see relation (4.26)] the potential energy of the body is

$$E_p = \frac{1}{2}kx^2 \tag{5.12}$$

We notice that E_p vanishes at the equilibrium position ($x = 0$) while it obtains a maximum value at the extreme points ($x = \pm A$) of the oscillation.

From (5.11) and (5.12) we can find the total mechanical energy of the oscillating body:

$$E = E_k + E_p = \frac{1}{2}kA^2 \tag{5.13}$$

Notice that this quantity assumes a fixed value during the SHM, in accordance with the principle of conservation of mechanical energy.

5.4 Oscillations of a Mass-Spring System

Springs are often used as instruments for producing SHM. Before we see how this is done, let us say a few words about the force a spring exerts on a body connected to it.

The spring may be in one of the following three states:

1. In its *natural length*, which occurs when the spring is not subject to external forces. The spring then, in turn, does *not* exert a force on a body connected to it.
2. In *extension* by Δl relative to its natural length. The spring then has a tendency to return to its natural length; it thus exerts a force $\vec{F}$ *opposite* to the extension, of magnitude $k\Delta l$, where k is called the *spring constant*.
3. In *compression* by Δl relative to its natural length. The spring again tends to return to its natural length, thus it exerts a force $\vec{F}$ *opposite* to the compression, also of magnitude $k\Delta l$.

a. *Horizontal oscillation*

The body, of mass m, is connected to a spring of spring constant k and moves along the x-axis on a frictionless horizontal surface, as shown in Fig. 5.5. At the location $x = 0$ (point O in the figure) the spring has its natural length and thus exerts no force on m. At a position $x \neq 0$, the spring suffers deformation (extension if $x > 0$ or compression if $x < 0$) and exerts a force on m, given by

$$\vec{F}_k = -kx\hat{u}_x \equiv F_k\hat{u}_x \tag{5.14}$$

Fig. 5.5 Horizontal oscillation of a mass-spring system

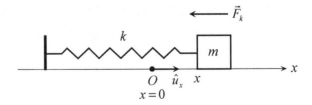

where $F_k = -kx$ is the algebraic value of $\vec{F}_k$. We note that $\vec{F}_k$ is directed left when m is to the right of O ($x > 0$), while its direction is to the right when m is to the left of O ($x < 0$). In any case, $\vec{F}_k$ is directed toward the equilibrium position O, at which position $\vec{F}_k$ vanishes.

Let us call the *resultant* force $\vec{F}$ on m. Clearly, $\vec{F} = \vec{F}_k$, since the other forces, namely, the weight of the body and the normal reaction from the horizontal surface, cancel each other (remember that there is no friction). Hence,

$$
\vec{F} = -kx\hat{u}_x \equiv F\hat{u}_x \Rightarrow
$$
$$
F = -kx
$$
(5.15)

where F is the algebraic value of $\vec{F}$. The fact that the *total* force on m is a *restoring force* of the form (5.15) indicates that the motion of the body is a SHM along the x-axis, centered at O [cf. Eq. (5.8)]. The angular frequency, the period and the frequency of oscillation are given by (5.9) and (5.10):

$$
\omega = \sqrt{\frac{k}{m}}, \quad T = \frac{1}{f} = 2\pi\sqrt{\frac{m}{k}}
$$
(5.16)

Note that these quantities do not depend on the amplitude of the SHM.

The potential energy of the body at the position x is

$$
E_p = \frac{1}{2}kx^2
$$
(5.17)

If A is the amplitude of oscillation, the total mechanical energy of the body is equal to

$$
E = E_k + E_p = \frac{1}{2}kA^2
$$
(5.18)

This quantity assumes a constant value in the course of the SHM. (Would this be the case if friction were present?)

b. *Vertical oscillation*

The spring is initially free and has its natural length l_0 (see Fig. 5.6). We then attach to the spring a body of mass m. When the body is *in equilibrium* at the position $x = 0$ of the vertical x-axis, the spring is extended by Δl and exerts on the body an upward vertical force equal to $F'_k = k\Delta l$ which force balances the weight mg of the body:

$$
k\Delta l = mg
$$
(5.19)

We now displace the body a distance x above the equilibrium position (that is, we raise the body from its initial position $x = 0$ to some position $x > 0$). The extension of the spring is now $(\Delta l - x)$ and the upward force on the body by the spring is F_k

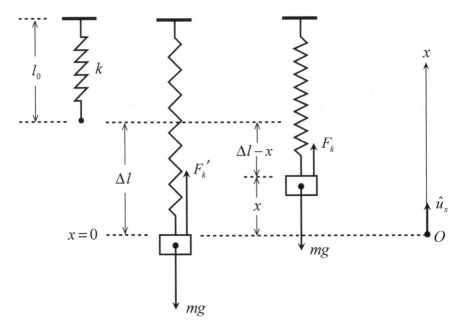

Fig. 5.6 Vertical oscillation of a mass-spring system

$= k\,(\Delta l - x)$. The *total* force on the body is

$$F = F_k - mg = k(\Delta l - x) - mg \quad \text{(algebraic value)} \ .$$

By making use of the equilibrium condition (5.19), we find:

$$F = -kx \tag{5.20}$$

Equation (5.20) was derived for $x > 0$, i.e., for a displacement of m *above* the equilibrium position. Consider now a displacement *below* the equilibrium position, at $x < 0$. The extension of the spring is then $\Delta l + |x| = \Delta l - x$, as before, and our previous expressions for the upward force F_k by the spring and for the total force F on m are still valid:

$$F_k = k(\Delta l + |x|) = k(\Delta l - x)$$
$$F = F_k - mg = k(\Delta l - x) - mg = -kx$$

where the equilibrium condition (5.19) was again used.

In conclusion, the *total* force on the body is a restoring force of the form (5.20), where x is the displacement from the equilibrium position. Under the action of this force the body will execute SHM *about its equilibrium position* (at which position the extension of the spring is Δl). The angular frequency and the period of oscillation are

$$\omega = \sqrt{\frac{k}{m}}, \quad T = \frac{1}{f} = 2\pi\sqrt{\frac{m}{k}} \tag{5.21}$$

The potential energy of the body at a position x is

$$E_p = \frac{1}{2}kx^2 \tag{5.22}$$

while the total mechanical energy of m is

$$E = E_k + E_p = \frac{1}{2}kA^2 \tag{5.23}$$

where A is the amplitude of oscillation. We note again that E is a constant quantity.

The force $\vec{F}_k$ exerted by a spring is conservative and is associated with a potential energy $E_{p,k}(y)$, where y is the deformation (extension or compression) of the spring. Indeed, this force is algebraically equal to

$$F_k = -ky$$

where we assume that $y > 0$ for extension and $y < 0$ for compression. As in Example 2 of Sect. 4.6, the work of $\vec{F}_k$ for a displacement of the point of application of this force from point A to point B is

$$W = \int_A^B \vec{F}_k \cdot d\vec{r} = -\int_A^B kydy = \frac{1}{2}ky_A^2 - \frac{1}{2}ky_B^2 \equiv E_{p,k}(A) - E_{p,k}(B)$$

and therefore,

$$E_{p,k}(y) = \frac{1}{2}ky^2 \tag{5.24}$$

where we have assumed that the potential energy vanishes for zero deformation of the spring.

By using (5.24) we can derive the expression (5.22) for the potential energy in a vertical oscillation, in the following alternative way: The *total* potential energy E_p of the oscillating body m is the sum of the potential energy $E_{p,m}$ due to gravity and the potential energy $E_{p,k}$ due to the deformation of the spring. By arbitrarily assuming that $E_{p,m}$ vanishes at the equilibrium position $x = 0$, and by using Eq. (4.22) of Chap. 4 with x in place of y, we have:

$$E_{p,m}(x) = mgx .$$

The deformation (extension) of the spring is $y = \Delta l - x$, so that, by (5.24),

$$E_{p,k}(x) = \frac{1}{2}k(\Delta l - x)^2 \ .$$

The total potential energy of m is

$$E_p(x) = E_{p,m}(x) + E_{p,k}(x) = mgx + \frac{1}{2}k(\Delta l - x)^2 = (mg - k\Delta l)x + \frac{1}{2}kx^2 + \frac{1}{2}k(\Delta l)^2 \ .$$

But, by the equilibrium condition (5.19), $mg - k\Delta l = 0$. Also, the term $k\,(\Delta l)^2/2$ is a constant quantity, independent of x, which may be omitted from the expression for the potential energy. Thus, finally,

$$E_p = E_{p,m} + E_{p,k} = \frac{1}{2}kx^2 \ ,$$

in agreement with (5.22).

5.5   Oscillation of a Pendulum

The oscillation of a pendulum follows *approximately* the laws of the SHM for small angles of deflection of the string from the vertical.

The mass m in Fig. 5.7 oscillates symmetrically about the equilibrium position O. For small angles θ of deflection, the arc OA may be approximated by a horizontal rectilinear segment of length $s = l\theta$, where l is the length of the pendulum string and where θ is in *rad*. Here, s and θ represent the displacement from the equilibrium position. In particular, $s > 0$ and $\theta > 0$ when m is to the right of O (in accordance with the positive direction of moving away from O, as defined by the unit tangent vector) while $s < 0$ and $\theta < 0$ when m is to the left of O. The restoring force that is responsible for the oscillation is the *tangential* component of the total force on m, that is, the component parallel to the velocity of m.

Fig. 5.7 Oscillation of a
pendulum

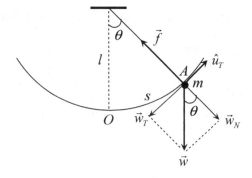

The mass m is subject to two forces; namely, the weight $\vec{w}$, of magnitude mg, and the tension $\vec{f}$ of the string. The total force on m is $\vec{F} = \vec{w} + \vec{f}$. As is easy to see, at any point A the tangential component $\vec{F}_T$ of $\vec{F}$ is equal to the tangential component of the weight:

$$\vec{F}_T = \vec{w}_T = -mg \sin \theta \, \hat{u}_T \equiv F_T \hat{u}_T$$

where $F_T = -mg \sin \theta$ is the algebraic value of $\vec{F}_T$. For sufficiently small values of the angle θ, we can make the approximation $\sin \theta \simeq \theta$. Hence,

$$F_T \simeq -mg\theta = -\left(\frac{mg}{l}\right) s \qquad (5.25)$$

(since $s = l\theta$). Relation (5.25) is of the form (5.8), with s in place of x:

$$F_T \simeq -ks, \quad \text{where} \quad k = \frac{mg}{l} .$$

This means that, under the action of $\vec{F}_T$, the mass m executes SHM about the equilibrium position O. The angular frequency and the period of oscillation are

$$\omega = \sqrt{\frac{k}{m}} = \sqrt{\frac{g}{l}}, \quad T = 2\pi\sqrt{\frac{m}{k}} = 2\pi\sqrt{\frac{l}{g}} \qquad (5.26)$$

Note that ω should *not* be interpreted as the angular velocity of the circular motion of m for *finite* values of θ and s. Indeed, since that motion is not uniform (why?) the corresponding angular velocity cannot be a constant, in contrast to the angular frequency ω in (5.26).

5.6 Differential Equation of SHM

As defined in Sect. 5.1, SHM is a special kind of rectilinear motion in which the displacement x from the equilibrium position O is given as a function of time by Eq. (5.1). By Newton's law, then, the *total* force on the oscillating body is a *restoring force* proportional to the displacement [see Eq. (5.8)].

Conversely, let us assume that a particle of mass m is subject to a total force of the form (5.8): $F = -kx$. According to Newton's law, this force is equal to $F = ma$, where

$$a = \frac{dv}{dt} = \frac{d}{dt}\left(\frac{dx}{dt}\right) = \frac{d^2x}{dt^2} .$$

We thus have:

$$m \frac{d^2x}{dt^2} = -kx \quad \Leftrightarrow \quad \frac{d^2x}{dt^2} + \frac{k}{m}x = 0 \; .$$

Setting $\omega^2 = k/m$ [comp. Eq. (5.9)] we finally have:

$$\frac{d^2x}{dt^2} + \omega^2 x = 0 \tag{5.27}$$

Relation (5.27) is a *second-order differential equation* for the function $x = x(t)$. Its *general solution* must depend on *two* arbitrary constants (or parameters). As can be verified by direct substitution into (5.27), this solution is precisely the function

$$x = A \cos (\omega t + \alpha)$$

by which the SHM was defined in Sect. 5.1. The amplitude A and the initial phase α are the two parameters which the solution of (5.27) is required to contain.

In the case of the pendulum (Sect. 5.5) the role of x is played by the arc $s = l\theta$, which arc is almost rectilinear for small angles θ. Given that

$$\frac{d^2s}{dt^2} = l \frac{d^2\theta}{dt^2} \; (\text{since } l \text{ is constant}),$$

the differential Eq. (5.27) with s in place of x reduces to

$$\frac{d^2\theta}{dt^2} + \omega^2 \theta = 0 \tag{5.28}$$

where $\omega^2 = g/l$ [see Eq. (5.26)]. The general solution of (5.28) has the form

$$\theta = \theta_0 \cos (\omega t + \alpha) \; ,$$

where θ_0 is the maximum angle of deflection of the string from the vertical.

In the real world, oscillations are not as simple as those described by (5.27). Thus, in addition to a restoring force of the form (5.8) the body may be subject to a frictional force and/or an applied periodic force. These physical situations are referred to as *damped oscillations* and *forced oscillations*, accordingly [1–3] and are described mathematically by linear (homogeneous or non-homogeneous, as the case may be) differential equations of the second order [4].

References

1. K.R. Symon, *Mechanics*, 3rd edn. (Addison-Wesley, 1971)
2. J.B. Marion, S.T. Thornton, *Classical Dynamics of Particles and Systems*, 4th edn. (Saunders College, 1995)
3. J.R. Taylor, *Classical Mechanics* (University Science Books, 2005)
4. A.F. Bermant, I.G. Aramanovich, *Mathematical Analysis* (Mir Publishers, 1975)

Chapter 6
Systems of Particles

Abstract The center of mass of a system of particles is defined and shown to play an important role in the dynamics of systems, in general. Conservation of momentum, angular momentum and mechanical energy of a system is examined. Elastic and plastic collisions are studied.

6.1 Center of Mass of a System of Particles

The dynamics of a system of particles presents additional difficulties compared to the single-particle case. What makes things more complicated is the fact that, in the case of a system one must distinguish between two kinds of forces; namely, *internal forces*—those exerted between particles of the system—and *external forces* resulting from factors not belonging to the system. As we will see, given a system one may find a certain point of space, called the *center of mass* of the system, which moves *as if* it were a particle of mass equal to the total mass of the system and subject to the total *external* force acting on the system.

Consider a system of particles of masses m_1, m_2, m_3, ... (Fig. 6.1). Assume that, at some particular moment, the particles are located at the points of space with corresponding position vectors $\vec{r}_1$, $\vec{r}_2$, $\vec{r}_3$, ..., relative to a reference point O which is typically chosen to be the origin of an inertial frame of reference.

The total mass of the system is

$$M = m_1 + m_2 + m_3 + \cdots = \sum_i m_i \tag{6.1}$$

The *center of mass* of the system is defined as the point C of space having position vector given by the equation

$$\vec{r}_C = \frac{1}{M} (m_1 \vec{r}_1 + m_2 \vec{r}_2 + \cdots) = \frac{1}{M} \sum_i m_i \vec{r}_i \tag{6.2}$$

C. J. Papachristou, *Introduction to Mechanics of Particles and Systems*,
https://doi.org/10.1007/978-3-030-54271-9_6

83

Fig. 6.1 A system of
particles and its center of
mass, C

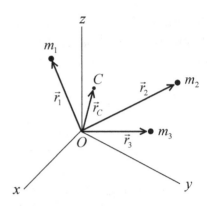

In relation (6.2) the position vectors of the particles and of the center of mass are
defined with respect to the fixed origin O of our coordinate system. If we choose a
different reference point O', these position vectors will, of course, change. However,
as will be shown in Appendices A and B, the position of the center of mass C *relative
to the system of particles* will remain the same, regardless of the choice of reference
point.

If (x_i, y_i, z_i) and (x_C, y_C, z_C) are the coordinates of m_i and C, respectively, we can
replace the vector relation (6.2) with three scalar equations:

$$x_C = \frac{1}{M} \sum_i m_i x_i \ , \quad y_C = \frac{1}{M} \sum_i m_i y_i \ , \quad z_C = \frac{1}{M} \sum_i m_i z_i \quad (6.3)$$

As an example, consider two particles of masses $m_1 = m$ and $m_2 = 2\,m$, located at
points x_1 and x_2 of the x-axis (Fig. 6.2). Call $a = x_2 - x_1$ the distance between these
particles. The total mass of the system is $M = m_1 + m_2 = 3\,m$. From relations (6.3)
it follows that the center of mass C of the system is located on the x-axis. Indeed,
$y_i = z_i = 0$ ($i = 1, 2$) so that $y_C = z_C = 0$ (the y and z-axes have not been drawn).
Furthermore,

$$x_C = \frac{1}{M} (m_1 x_1 + m_2 x_2) = \frac{1}{3} (x_1 + 2x_2) = x_1 + \frac{2}{3} a$$

where we have used the fact that $x_2 = x_1 + a$. Thus, the center of mass C is located at
a distance $2a/3$ from m. Notice that the position of C *relative to the system of particles*

Fig. 6.2 A system of two
masses m and $2\,m$

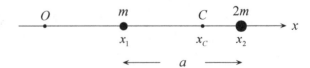

does not depend on the choice of the reference point O with respect to which the coordinates of the particles are determined.

As the above example demonstrates, the position of the center of mass does not necessarily coincide with the position of a particle of the system. (Give examples of systems in which a particle is located at C, as well as of systems where no such coincidence occurs.).

6.2 Newton's Second Law and Conservation of Momentum

Let $\vec{v}_1$, $\vec{v}_2$,..., be the instantaneous velocities of the particles m_1, m_2, ..., of a system, at time t. The *total momentum* of the system at this time is

$$\vec{P} = \vec{p}_1 + \vec{p}_2 + \cdots = m_1\vec{v}_1 + m_2\vec{v}_2 + \cdots$$

or, briefly,

$$\vec{P} = \sum_i \vec{p}_i = \sum_i m_i \vec{v}_i \qquad (6.4)$$

For simplicity, we consider a system of two particles m_1, m_2 (Fig. 6.3). We call $\vec{F}_1$, $\vec{F}_2$ the corresponding *external* forces on these particles, due to the presence of other bodies not belonging to the system (e.g., the weights of m_1, m_2, due to the attraction of the Earth). We assume further that *internal* forces are also present; namely, $\vec{F}_{12}$ on particle 1 due to its interaction with particle 2, and $\vec{F}_{21}$ on particle 2 due to particle 1. According to Newton's third law, $\vec{F}_{12} = -\vec{F}_{21}$.

The total force on particle 1 is $\vec{F}_1 + \vec{F}_{12}$, while that on particle 2 is $\vec{F}_2 + \vec{F}_{21}$. By Newton's second law,

$$\frac{d\vec{p}_1}{dt} = \vec{F}_1 + \vec{F}_{12} , \qquad \frac{d\vec{p}_2}{dt} = \vec{F}_2 + \vec{F}_{21}$$

Adding the above equations and taking into account that $\vec{F}_{12} + \vec{F}_{21} = 0$, we find:

$$\frac{d\vec{p}_1}{dt} + \frac{d\vec{p}_2}{dt} = \frac{d}{dt}(\vec{p}_1 + \vec{p}_2) = \vec{F}_1 + \vec{F}_2$$

Fig. 6.3 Internal and external forces on a system of two particles

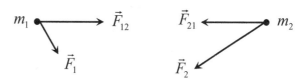

Generalizing the above result for a system with an arbitrary number of particles [1, 2] we have:

$$\frac{d}{dt} \sum_i \vec{p}_i = \sum_i \vec{F}_i \qquad (6.5)$$

The left-hand side of (6.5) is the time derivative of the total momentum $\vec{P}$ of the system. The right-hand side represents the *total external force* acting on the system:

$$\vec{F}_{\text{ext}} = \sum_i \vec{F}_i \qquad (6.6)$$

Thus, (6.5) is written:

$$\boxed{\frac{d\vec{P}}{dt} = \vec{F}_{\text{ext}}} \qquad (6.7)$$

That is,

the rate of change of the total momentum of the system equals the total external force acting on the system.

Note that *internal* forces *cannot* by themselves alter the total momentum of the system; for a change of total momentum, *external* forces are needed.

How is the center of mass related to all this? The answer is contained in the following two propositions:

1. *The total momentum of the system is equal to the momentum of a hypothetical particle having mass equal to the total mass of the system and moving with the velocity of the center of mass of the system.*
2. *The equation of motion of the center of mass of the system is that of a hypothetical particle of mass equal to the total mass of the system, subject to the total external force acting on the system.*

Proof

1. By differentiating (6.2) with respect to time, we find the velocity of the center of mass of the system:

$$\vec{v}_C = \frac{d\vec{r}_C}{dt} = \frac{d}{dt}\left(\frac{1}{M}\sum_i m_i \vec{r}_i\right) = \frac{1}{M}\sum_i m_i \frac{d\vec{r}_i}{dt}$$

$$\Rightarrow \vec{v}_C = \frac{1}{M}\sum_i m_i \vec{v}_i = \frac{1}{M}\sum_i \vec{p}_i \qquad (6.8)$$

Combining this with (6.4), we have:

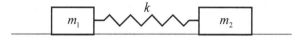

Fig. 6.4 Two masses connected by a spring and moving on a frictionless horizontal plane

$$\boxed{\vec{P} = M\,\vec{v}_C}$$ (6.9)

2. Differentiating (6.9), we have:

$$\frac{d\vec{P}}{dt} = \frac{d}{dt}\,(M\,\vec{v}_C) = M\,\frac{d\vec{v}_C}{dt} = M\,\vec{a}_C$$

where $\vec{a}_C$ is the acceleration of the center of mass. Hence, by (6.7),

$$\boxed{\vec{F}_{\text{ext}} = M\,\vec{a}_C}$$ (6.10)

Strictly speaking, a system of particles is *isolated* if it is not subject to any external interactions (a situation that is only theoretically possible). More generally, we will say that a system is "macroscopically isolated" or, simply, "isolated" if the total external force on it is zero: $\vec{F}_{\text{ext}} = 0$. In this case, relations (6.7) and (6.9) lead to the following conclusions:

1. *The total momentum of an isolated system of particles retains a constant value relative to an inertial frame of reference (principle of conservation of momentum).*
2. *The center of mass of an isolated system of particles moves with constant velocity relative to an inertial reference frame.*

As an example, consider two masses m_1 and m_2 connected to each other by a spring. The masses can move on a frictionless horizontal plane, as shown in Fig. 6.4. The system may be considered isolated since the total external force on it is zero (explain this). Thus, the total momentum of the system and the velocity of the center of mass C remain constant while the two masses move on the plane. Note that the *internal* force $F_{\text{int}} = k\,\Delta l$, where Δl is the deformation of the spring relative to its natural length, *cannot* produce any change to the total momentum and the velocity of C.

6.3 Angular Momentum of a System of Particles

As noted in Sect. 3.7, the angular momentum of a particle is not an absolute quantity (such as, e.g., the momentum) but is always defined relative to some reference point O. We choose O to be the origin of coordinates of an inertial reference frame (we recall that in such frames only the laws of Newton are valid).

Given a system of particles, the angular momentum of a particle m_i is

$$\vec{L}_i = \vec{r}_i \times \vec{p}_i = m_i \left(\vec{r}_i \times \vec{v}_i \right) \tag{6.11}$$

where $\vec{r}_i$ is the position vector of m_i relative to O. According to (3.38), if $\vec{F}_{i,\text{tot}}$ is the total force on m_i, the torque of $\vec{F}_{i,\text{tot}}$ relative to O is

$$\vec{T}_i = \vec{r}_i \times \vec{F}_{i,\text{tot}} = \frac{d\vec{L}_i}{dt} \tag{6.12}$$

Let us initially consider a system of two particles m_1 and m_2 that are subject to external forces $\vec{F}_1$ and $\vec{F}_2$, respectively, while the corresponding internal forces are, in the notation of Sect. 6.2, $\vec{F}_{12}$ and $\vec{F}_{21}$, with $\vec{F}_{12} = -\vec{F}_{21}$ as required by the action-reaction law. By (6.12),

$$\frac{d\vec{L}_1}{dt} = \vec{T}_1 \;, \qquad \frac{d\vec{L}_2}{dt} = \vec{T}_2$$

and, by adding these,

$$\frac{d\vec{L}_1}{dt} + \frac{d\vec{L}_2}{dt} = \frac{d}{dt} \left(\vec{L}_1 + \vec{L}_2 \right) = \vec{T}_1 + \vec{T}_2 \tag{6.13}$$

But,

$$\vec{T}_1 = \vec{r}_1 \times \vec{F}_{1,\text{tot}} = \vec{r}_1 \times \left(\vec{F}_1 + \vec{F}_{12} \right) = \vec{r}_1 \times \vec{F}_1 + \vec{r}_1 \times \vec{F}_{12} \;,$$
$$\vec{T}_2 = \vec{r}_2 \times \vec{F}_{2,\text{tot}} = \vec{r}_2 \times \left(\vec{F}_2 + \vec{F}_{21} \right) = \vec{r}_2 \times \vec{F}_2 + \vec{r}_2 \times \vec{F}_{21} \;.$$

Adding the above equations and taking into account that $\vec{F}_{12} = -\vec{F}_{21}$, we have:

$$\vec{T}_1 + \vec{T}_2 = \vec{r}_1 \times \vec{F}_1 + \vec{r}_2 \times \vec{F}_2 + \left(\vec{r}_1 - \vec{r}_2 \right) \times \vec{F}_{12}$$

The difference $\left(\vec{r}_1 - \vec{r}_2 \right)$ is a vector directed along the line joining the two particles (see Sect. 1.3). If we make the *additional* assumption that the internal forces $\vec{F}_{12}$ and $\vec{F}_{21}$ act along this line, then

$$\left(\vec{r}_1 - \vec{r}_2 \right) \times \vec{F}_{12} = 0$$

so that

$$\vec{T}_1 + \vec{T}_2 = \vec{r}_1 \times \vec{F}_1 + \vec{r}_2 \times \vec{F}_2 = \vec{T}_{\text{ext}}$$

where $\vec{T}_{\text{ext}}$ is the *total external torque* on the system of particles, relative to O. Equation (6.13) is thus written:

$$\frac{d}{dt}(\vec{L}_1 + \vec{L}_2) = \vec{r}_1 \times \vec{F}_1 + \vec{r}_2 \times \vec{F}_2 = \vec{T}_{\text{ext}} \tag{6.14}$$

For a system with an arbitrary number of particles, (6.14) is generalized as follows [2]:

$$\frac{d}{dt}\sum_i \vec{L}_i = \sum_i \vec{r}_i \times \vec{F}_i = \vec{T}_{\text{ext}} \tag{6.15}$$

The vector

$$\vec{L} = \sum_i \vec{L}_i = \vec{L}_1 + \vec{L}_2 + \cdots \tag{6.16}$$

represents the *total angular momentum* of the system relative to O. By (6.15),

$$\boxed{\frac{d\vec{L}}{dt} = \vec{T}_{\text{ext}}} \tag{6.17}$$

That is,

the rate of change of the total angular momentum of the system, relative to O, equals the total external torque on the system with respect to O.

We note that the *internal* forces alone *cannot* produce a change to the total angular momentum of the system. [Notice the similarity in form between (6.7) and (6.17).]

Given that (6.17) is a consequence of Newton's law, this relation is expected to be valid in inertial reference frames only. Thus, the point O relative to which $\vec{L}$ and $\vec{T}_{\text{ext}}$ are evaluated must be the origin of some inertial frame. For an isolated system of particles, the center of mass C moves with constant velocity with respect to any inertial frame (see Sect. 6.2) and therefore C itself may be considered as the origin of a special inertial frame, called the *center-of-mass frame* (or briefly, *C-frame*). Thus, for an isolated system of particles, relation (6.17) is valid if $\vec{L}$ and $\vec{T}_{\text{ext}}$ are taken with respect to the center of mass C.

When the system of particles is not isolated, its center of mass C is accelerating according to (6.10). Thus, C *cannot* be the origin of some inertial reference frame. One might therefore assume that, in this case, relation (6.17) is not valid with respect to C. One of the curious facts of Mechanics, however, is that this latter relation *is* valid relative to C even if the system is *not* isolated (see Appendix B for a proof). That is,

relation (6.17) is always valid relative to the center of mass C of the system, even if C is accelerating with respect to an inertial observer.

It follows from (6.17) that, if $\vec{T}_{\text{ext}} = 0$, then $d\vec{L}/dt = 0 \Leftrightarrow \vec{L} = \text{constant}$. This observation leads us to the *principle of conservation of angular momentum*:

When the total external torque on a system, relative to some point O, is zero, the total angular momentum of the system with respect to this point is constant in time.

As an example, consider a system of electrically charged particles with charges $q_1, q_2, \ldots$ The particles are located inside the electric field created by some external charge Q firmly placed at a point O. No other external forces other than the Coulomb forces from Q are exerted on the system. Assuming that the relative velocities of the charges are sufficiently small, in order for the electromagnetic force between any two charges to be approximately directed along the line joining these charges, show that the total angular momentum of the system of the q_i, relative to O, is constant in time. (Note that this statement is generally *not* true for reference points other than the point O at which the external charge Q is located, since, relative to such points, $\vec{T}_{\text{ext}} \neq 0$.)

Comment: The principle of conservation of angular momentum was based on (6.17) which, in turn, was derived by assuming that (*a*) the action-reaction law (3.6) is valid, and (*b*) the internal forces of the system are directed along the lines joining the particles of the system (that is, the internal forces are *central* forces). There are, however, physical situations in which these conditions are not fulfilled (this happens, for example, in the case of a system of moving electric charges the relative velocities of which are not small). Relations (6.7) and (6.17) are *not* valid for such systems and conservation of momentum and angular momentum is not satisfied, even in the absence of external forces. Conservation laws are restored by assuming that, in addition to the momentum and the angular momentum of the system of charges, one must take into account the corresponding quantities contained in the electromagnetic field itself. (Yes, even a field may carry energy, momentum and angular momentum! See, e.g., [3].) We thus see that the conservation laws we have found have an even deeper and more general meaning.

Note: Let O be the origin of coordinates of an inertial reference frame, and let C be the center of mass of a system of particles. As shown in Appendix B, the angular momentum $\vec{L}$ of the system relative to O, and the angular momentum $\vec{L}_C$ of the system relative to C, are related by

$$\vec{L} = \vec{L}_C + M\,(\vec{r}_C \times \vec{v}_C),$$

where M is the total mass of the system, and where $\vec{r}_C$ and $\vec{v}_C$ are the position vector and the velocity, respectively, of the center of mass C relative to O.

6.4   Kinetic Energy of a System of Particles

According to the work-energy theorem (Sect. 4.3) the change of the kinetic energy of a particle, within a period of time, equals the work of the *total* force acting on the particle within that period. For a system of particles, the work-energy theorem is generalized as follows:

The change of the total kinetic energy of a system of particles is equal to the work done by the external *and* the internal forces acting on the system.

Proof Consider, for simplicity, a system of two particles of masses m_1, m_2, subject to external forces $\vec{F}_1$, $\vec{F}_2$ and internal forces $\vec{F}_{12}$, $\vec{F}_{21}$, respectively. Let $d\vec{r}_1$, $d\vec{r}_2$ be the elementary displacements of the particles in time dt. By Newton's law,

$$m_1\vec{a}_1 = \vec{F}_1 + \vec{F}_{12}, \qquad m_2\vec{a}_2 = \vec{F}_2 + \vec{F}_{21}$$

(where $\vec{a}_1$, $\vec{a}_2$ are the accelerations of the particles), so that

$$m_1\vec{a}_1 \cdot d\vec{r}_1 = \vec{F}_1 \cdot d\vec{r}_1 + \vec{F}_{12} \cdot d\vec{r}_1 \ ,$$
$$m_2\vec{a}_2 \cdot d\vec{r}_2 = \vec{F}_2 \cdot d\vec{r}_2 + \vec{F}_{21} \cdot d\vec{r}_2 \ .$$

Adding these and taking into account that $\vec{F}_{21} = -\vec{F}_{12}$, we have:

$$m_1\vec{a}_1 \cdot d\vec{r}_1 + m_2\vec{a}_2 \cdot d\vec{r}_2 = \vec{F}_1 \cdot d\vec{r}_1 + \vec{F}_2 \cdot d\vec{r}_2 + \vec{F}_{12} \cdot (d\vec{r}_1 - d\vec{r}_2) \qquad (6.18)$$

But,

$$\vec{a}_1 \cdot d\vec{r}_1 = \frac{d\vec{v}_1}{dt} \cdot d\vec{r}_1 = d\vec{v}_1 \cdot \frac{d\vec{r}_1}{dt} = \vec{v}_1 \cdot d\vec{v}_1 = \frac{1}{2} d(\vec{v}_1 \cdot \vec{v}_1) = \frac{1}{2} d(v_1^2)$$
$$= \frac{1}{2} \frac{d(v_1^2)}{dv_1} dv_1 = \frac{1}{2} (2v_1) dv_1 = v_1 dv_1$$

and similarly,

$$\vec{a}_2 \cdot d\vec{r}_2 = v_2 dv_2$$

where v_1, v_2 are the magnitudes of $\vec{v}_1$, $\vec{v}_2$, respectively. Furthermore,

$$d\vec{r}_1 - d\vec{r}_2 = d(\vec{r}_1 - \vec{r}_2) = d\vec{r}_{12}$$

where $\vec{r}_{12} \equiv \vec{r}_1 - \vec{r}_2$. Hence, (6.18) is written:

$$m_1 v_1 dv_1 + m_2 v_2 dv_2 = \vec{F}_1 \cdot d\vec{r}_1 + \vec{F}_2 \cdot d\vec{r}_2 + \vec{F}_{12} \cdot d\vec{r}_{12}$$

Integrating the above differential relation from A to B, where by A and B we denote the state of the system (e.g., the locations and velocities of the particles) at times t_A and t_B, respectively, we have:

$$m_1 \int_A^B v_1 dv_1 + m_2 \int_A^B v_2 dv_2 = \int_A^B \vec{F}_1 \cdot d\vec{r}_1 + \int_A^B \vec{F}_2 \cdot d\vec{r}_2 + \int_A^B \vec{F}_{12} \cdot d\vec{r}_{12}$$

$$(6.19)$$

The left-hand side of (6.19) is equal to

$$m_1 \left[\frac{v_1^2}{2}\right]_A^B + m_2 \left[\frac{v_2^2}{2}\right]_A^B = \left(\frac{1}{2}m_1 v_1^2 + \frac{1}{2}m_2 v_2^2\right)_B - \left(\frac{1}{2}m_1 v_1^2 + \frac{1}{2}m_2 v_2^2\right)_A$$
$$= E_{k,B} - E_{k,A} \equiv \Delta E_k$$

where the quantity

$$E_k = \frac{1}{2}m_1 v_1^2 + \frac{1}{2}m_2 v_2^2 = \sum_i \frac{1}{2}m_i v_i^2 \qquad (6.20)$$

represents the *total kinetic energy* of the system. The sum

$$\int_A^B \vec{F}_1 \cdot d\vec{r}_1 + \int_A^B \vec{F}_2 \cdot d\vec{r}_2 = \sum_i \int_A^B \vec{F}_i \cdot d\vec{r}_i = W_{\text{ext}} \qquad (6.21)$$

on the right-hand side of (6.19) represents the total work of the *external* forces in the period between t_A and t_B, while the integral

$$\int_A^B \vec{F}_{12} \cdot d\vec{r}_{12} = \int_A^B \vec{F}_{12} \cdot d\vec{r}_1 + \int_A^B \vec{F}_{21} \cdot d\vec{r}_2 = \sum_i \sum_{j \neq i} \int_A^B \vec{F}_{ij} \cdot d\vec{r}_i = W_{\text{int}}$$
$$(6.22)$$

represents the work of the *internal* forces in that period. Thus, finally, (6.19) is written:

$$\boxed{E_{k,B} - E_{k,A} \equiv \Delta E_k = W_{\text{ext}} + W_{\text{int}}} \qquad (6.23)$$

Although proven for a two-particle system, relation (6.23) is generally valid for a system with any number of particles [2].

If no external forces act on the system, then $W_{\text{ext}} = 0$ and any change of the total kinetic energy is due exclusively to the *internal* forces. As an example, consider two electric charges that are initially kept at rest. If they are allowed to move freely, their mutual electrical interaction will set the charges in motion and the system will acquire kinetic energy equal to the work of the internal Coulomb forces. Note that, in the absence of *external* forces, the total momentum of a system is constant, which is generally *not* the case with regard to the kinetic energy (unless the particles interact very weakly with one another, so that internal forces may be ignored).

Note: As shown in Appendix B, the kinetic energy E_k of a system of particles relative to the origin O of our inertial frame of reference, and the kinetic energy $E_{k,C}$ of the system relative to its center of mass C, are related by

$$E_k = E_{k,C} + \frac{1}{2}M\,v_C^2$$

where M is the total mass of the system and where $\vec{v}_C$ is the velocity of the center of mass C with respect to O.

6.5   Total Mechanical Energy of a System of Particles

The work-energy theorem (6.23) is of general validity, independently of the nature of the external and the internal forces acting on a system of particles. In (6.23), W_{ext} and W_{int} represent the corresponding works of these forces when the system progresses from a state A at time t_A to a state B at time t_B.

When both the internal and the external forces are *conservative*, one may find an *internal potential energy* $E_{p,\,int}$ and an *external potential energy* $E_{p,\,ext}$, both functions of the coordinates of the particles, such that

$$W_{int} = (E_{p,int})_A - (E_{p,int})_B \tag{6.24}$$

and

$$W_{ext} = (E_{p,ext})_A - (E_{p,ext})_B \tag{6.25}$$

By (6.23) we then have:

$$E_{k,B} - E_{k,A} = (E_{p,int} + E_{p,ext})_A - (E_{p,int} + E_{p,ext})_B$$
$$\Rightarrow (E_k + E_{p,int} + E_{p,ext})_A = (E_k + E_{p,int} + E_{p,ext})_B \tag{6.26}$$

Relation (6.26), expressing *conservation of mechanical energy* for a system of particles, is valid for any two states A and B of the system. We conclude that the quantity

$$E = E_k + E_{p,int} + E_{p,ext} = E_k + E_p \tag{6.27}$$

representing the *total mechanical energy* of the system, is constant during the motion of the system when *all* forces acting on it—both internal and external—are conservative. Note that we have called $E_p = E_{p,\,int} + E_{p,\,ext}$ the *total potential energy* of the system. Note also that the expressions $E_{p,\,int}$ and $E_{p,\,ext}$ represent sums of potential energies associated with all conservative forces (both internal and external) acting on the system.

When some of the forces (internal and/or external) acting on the system are *non-conservative*, the sum $(E_k + E_p)$ is generally not constant: its change is given by a relation analogous to (4.20). Thus, if W' is the work of the non-conservative forces as the system progresses from a state A to a state B, it is not hard to show that

$$W' = (E_k + E_p)_B - (E_k + E_p)_A \equiv \Delta (E_k + E_p) \qquad (6.28)$$

That is,

the change of the sum $(E_k + E_p)$ equals the work of the non-conservative forces (both internal and external).[1]

The sum $(E_k + E_p)$ is constant only in the special case where the non-conservative forces do not produce work ($W' = 0$).

Let us see two characteristic examples of conservative systems:

1. Consider a hydrogen atom, consisting of an electron and a proton of masses m_1 and m_2, respectively. We assume that the particles are subject to no forces other than their mutual Coulomb attraction (4.28), which is a conservative internal force of the system (see Sect. 4.6). We have:

$$E_k = \frac{1}{2}m_1 v_1^2 + \frac{1}{2}m_2 v_2^2 , \qquad E_{p,\text{int}} = -k\frac{q^2}{r} , \qquad E_{p,\text{ext}} = 0$$

where q is the absolute value of the charge of the electron and where r is the distance of the electron from the nucleus of the atom (i.e., from the proton). Conservation of total mechanical energy is then expressed by the relation

$$E = \frac{1}{2}m_1 v_1^2 + \frac{1}{2}m_2 v_2^2 - k\frac{q^2}{r} = const. \qquad (6.29)$$

Exercise By using (6.22) and (6.24), together with (4.28), verify the expression given above for the internal potential energy $E_{p,\text{ int}}$.

2. Consider two masses m_1, m_2 connected to each other by a spring of spring constant k (Fig. 6.5). At a given moment the system is in the air near the surface of the Earth. We call y_1, y_2 the instantaneous heights at which the masses are located above a reference level $y = 0$ (e.g., the surface of the Earth) at which the gravitational potential energy is assumed to be zero, and we denote by x the deformation (extension or compression) of the spring relative to its natural length. There are two internal forces in the system, of magnitude kx and of opposite directions, which forces are due to the mutual interaction of the two masses through the spring. On the other hand, the external forces $m_1 g$ and $m_2 g$ are due to the gravitational attraction of the Earth on the two masses. The potential energies (internal and external) of the system are

$$E_{p,\text{int}} = \frac{1}{2}k x^2 , \qquad E_{p,\text{ext}} = m_1 g y_1 + m_2 g y_2$$

---

[1]For a more general interpretation of (6.28), see Note at the end of Sect. 4.5.

Fig. 6.5 Two masses
connected by a spring and
moving in the air near the
surface of the Earth

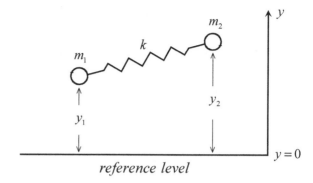

reference level

(Note that $E_{p,\,ext}$ is the sum of the gravitational potential energies associated with each mass separately.) By conservation of mechanical energy,

$$E = \frac{1}{2}m_1 v_1^2 + \frac{1}{2}m_2 v_2^2 + \frac{1}{2}k x^2 + m_1 g y_1 + m_2 g y_2 = \; const. \tag{6.30}$$

Exercise By using (6.22) and (6.24), verify the expression given above for the internal potential energy $E_{p,\,int.}$ (You may assume that the two masses always move along the line joining them.) Assume next that the masses move on a frictionless inclined plane. By using (6.28) and by taking into account that the normal reactions from the plane produce no work ($W' = 0$) conclude that (6.30) is still valid in this case.

6.6   Collisions

A *collision* is a form of interaction between two masses, during which the masses are momentarily "in contact" (though this never occurs at the atomic level!), exchanging momentum, angular momentum and energy in the process. A collision is assumed to take place within an extremely small (infinitesimal) time interval dt. That is, the interaction between the masses is almost instantaneous: just before and right after contact, the masses essentially exert no forces on each other.

Collision problems offer us a good opportunity to study various conservation laws in practice:

A.   *Conservation of momentum*

As we know, the change $d\vec{P}$ of total momentum $\vec{P}$ of a system of particles, within a time interval dt, is due to the total external force $\vec{F}_{ext}$ (e.g., gravity, friction, etc.) that acts on the system in that period. In mathematical terms,

$$\frac{d\vec{P}}{dt} = \vec{F}_{\text{ext}} \quad \Rightarrow \quad d\vec{P} = \vec{F}_{\text{ext}}\, dt$$

In a collision, the external forces are almost negligible compared to the internal forces that are associated with the mutual interaction of the two masses. Furthermore, the duration dt of this interaction is infinitesimal. We may thus make the approximation

$$\vec{F}_{\text{ext}}\, dt \approx 0 \quad \Rightarrow \quad d\vec{P} \approx 0$$

We conclude that

the total momentum of the system is unchanged in a collision; that is, the momentum *just before* the collision equals the momentum *right after* the collision.

B. *Conservation of angular momentum*

The change $d\vec{L}$ of total angular momentum of a system, relative to some point O, and the total external torque $\vec{T}_{\text{ext}}$ on the system, relative to O, are related by

$$\frac{d\vec{L}}{dt} = \vec{T}_{\text{ext}} \quad \Rightarrow \quad d\vec{L} = \vec{T}_{\text{ext}}\, dt$$

As before,

$$\vec{T}_{\text{ext}}\, dt \approx 0 \quad \Rightarrow \quad d\vec{L} \approx 0$$

That is,

in a collision, the total angular momentum of the system, relative to any point, is unchanged (i.e., the values of the angular momentum *just before* and *right after* the collision are the same).

C. *Kinetic energy*

In contrast to momentum and angular momentum, kinetic energy is *not* necessarily conserved in a collision. This is explained as follows: We recall that the change of total kinetic energy of a system, within a time period (here, the time dt of duration of the collision), equals the work of the external *and* the internal forces acting on the system in that period. Even if we assume that, for reasons explained previously, the work of the external forces within an infinitesimal time interval is negligible, we cannot ignore the work of the internal forces, which is particularly significant if the collision produces a *deformation* of the colliding bodies. In such a case, part of the initial kinetic energy of the system is lost due to the (negative) work done by the internal forces that are responsible for this deformation.

Collisions are classified on the basis of conservation or non-conservation of kinetic energy, as follows:

a. *Elastic collision*: The total kinetic energy of the system of colliding bodies is unchanged by the collision. That is, the kinetic energies *just before* and *right after* the collision are equal. This is the case for collisions that do not cause deformation of the colliding bodies. Example: the collision of two hard billiard balls.

b. *Inelastic collision*: Part of the kinetic energy of the system is lost due to deformation caused by the collision; this energy is thus not conserved in the process. Example: the collision of two rubber balls.

c. *Completely inelastic* (or *plastic*) *collision*: This is an extreme form of inelastic collision in which the colliding bodies stick together and move as one body, with common final velocity. Part of the initial kinetic energy of the system is lost due to deformation of the bodies (see Problem 37). Example: the collision of two lumps of wet clay.

Let us examine cases (*a*) and (*c*) in more detail:

Plastic collision

Consider two masses m_1 and m_2 moving along the x-axis with velocities $\vec{v}_1 = v_1 \hat{u}_x$ and $\vec{v}_2 = v_2 \hat{u}_x$, as seen in Fig. 6.6. (Note that v_1 and v_2 are *algebraic values* that may be positive or negative. In Fig. 6.6, $v_1 > 0$ and $v_2 < 0$.) After colliding, the masses stick to each other and move as one body of mass $(m_1 + m_2)$, with velocity $\vec{V} = V \hat{u}_x$.

As argued above, conservation of momentum (though *not* of kinetic energy) applies in this case. Equating the values of total momentum just before and right after the collision, we have:

$$m_1 \vec{v}_1 + m_2 \vec{v}_2 = (m_1 + m_2)\, \vec{V}$$

$$\Rightarrow \vec{V} = \frac{m_1 \vec{v}_1 + m_2 \vec{v}_2}{m_1 + m_2} = \frac{m_1 v_1 + m_2 v_2}{m_1 + m_2} \hat{u}_x \equiv V \hat{u}_x \qquad (6.31)$$

In the special case where the two masses are equal $(m_1 = m_2)$ and collide head-on with equal speeds $(v_2 = -v_1)$, Eq. (6.31) yields $\vec{V} = 0$. That is, the resulting composite mass stays at rest. In general, the direction of the velocity $\vec{V}$ of this mass is determined by the sign of the algebraic value V.

Exercise Find the velocity $\vec{V}$ for the case where the body m_2 is initially at rest. What will happen if $m_2 \gg m_1$ (as, e.g., when a lump of wet clay is ejected against a wall)?

Before   After

Fig. 6.6 Plastic collision of two masses

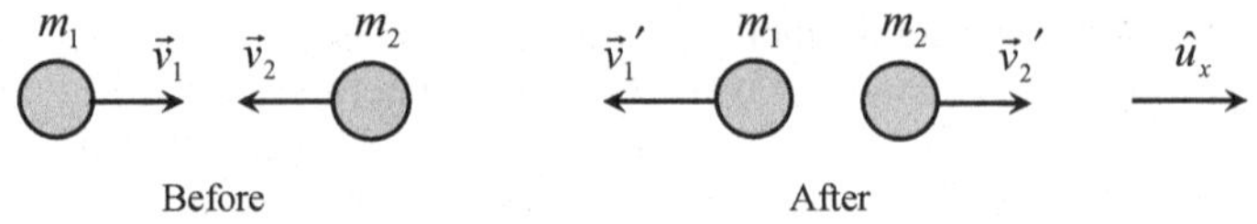

Fig. 6.7 Elastic collision of two masses

Elastic collision

Consider two masses m_1 and m_2 moving along the x-axis with velocities $\vec{v}_1 = v_1 \hat{u}_x$ and $\vec{v}_2 = v_2 \hat{u}_x$ (Fig. 6.7). After colliding elastically, the masses acquire new velocities $\vec{v}_1' = v_1' \hat{u}_x$ and $\vec{v}_2' = v_2' \hat{u}_x$. (The v_1, v_2, v_1', v_2' are algebraic values that may be positive or negative. In Fig. 6.7, $v_1 > 0$, $v_2 < 0$, etc.) We seek the velocities of the two masses after the collision.

By conservation of momentum,

$$m_1 \vec{v}_1 + m_2 \vec{v}_2 = m_1 \vec{v}_1' + m_2 \vec{v}_2' \quad \Rightarrow \quad (m_1 v_1 + m_2 v_2)\, \hat{u}_x$$
$$= (m_1 v_1' + m_2 v_2')\, \hat{u}_x \quad \Rightarrow m_1 v_1 + m_2 v_2 = m_1 v_1' + m_2 v_2' \quad (6.32)$$

Furthermore, since total kinetic energy is also conserved,

$$\frac{1}{2} m_1 v_1^2 + \frac{1}{2} m_2 v_2^2 = \frac{1}{2} m_1 v_1'^2 + \frac{1}{2} m_2 v_2'^2$$
$$\Rightarrow m_1 v_1^2 + m_2 v_2^2 = m_1 v_1'^2 + m_2 v_2'^2 \quad (6.33)$$

Equation (6.32) is written:

$$m_1 (v_1 - v_1') = m_2 (v_2' - v_2) \quad (6.32')$$

while (6.33) yields

$$m_1 (v_1 - v_1')(v_1 + v_1') = m_2 (v_2' - v_2)(v_2 + v_2') \quad (6.33')$$

We make the logical assumption that the velocities of the two bodies change as a result of the collision. Hence, $v_1 - v_1' \neq 0$, $v_2' - v_2 \neq 0$, which fact allows us to divide (6.33') by (6.32'):

$$v_1 + v_1' = v_2 + v_2' \quad (6.34)$$

Relations (6.32) and (6.34) are a system of equations with unknowns v_1' and v_2', for given v_1 and v_2. By solving this system, we find:

$$v_1' = \frac{(m_1 - m_2) v_1 + 2 m_2 v_2}{m_1 + m_2} \quad , \quad v_2' = \frac{2 m_1 v_1 + (m_2 - m_1) v_2}{m_1 + m_2} \quad (6.35)$$

Special cases:

1. If $m_1 = m_2$, Eq. (6.35) yields $v_1' = v_2$, $v_2' = v_1$. That is, *the two bodies exchange velocities*. In particular, if one of the bodies is initially at rest, then after the collision it acquires the initial velocity of the other body, which, in turn, comes to rest.
2. If $m_2 \gg m_1$, we can make the approximation $m_1/m_2 \simeq 0$. Equation (6.35) then yields $v_1' \simeq -v_1 + 2v_2$, $v_2' \simeq v_2$ (show this). In particular, if m_2 is initially at rest ($v_2 = 0$) then $v_1' \simeq -v_1$, $v_2' \simeq 0$. That is, after the collision the body m_2 remains at rest, while the direction of motion of m_1 is reversed with no change in this body's speed.

References

1. C.J. Papachristou, Foundations of newtonian dynamics: an axiomatic approach for the thinking student. Nausivios Chora **4**, 153 (2012). http://nausivios.snd.edu.gr/docs/partC2012.pdf; new version: https://arxiv.org/abs/1205.2326
2. J.B. Marion, S.T. Thornton, *Classical Dynamics of Particles and Systems*, 4th Edn. (Saunders College, 1995)
3. D.J. Griffiths, *Introduction to Electrodynamics*, 4th Edn. (Pearson, 2013)

Chapter 7
Rigid-Body Motion

Abstract The center of mass of a rigid body is defined. The concept of the moment of inertia is introduced and shown to be intimately related to angular momentum. The equations of motion (translational and rotational) of a rigid body are written. Conservation of angular momentum of a rigid body is studied, and the conditions for equilibrium of a body are examined. The rolling of a body on a plane surface is studied and the concept of gyroscopic motion is introduced.

7.1 Rigid Body

A system of particles constitutes a *rigid body* if the relative positions and distances of the particles remain fixed when external forces or torques act on the system. Hence, a rigid body maintains its shape during its motion.[1]

A rigid body may execute two kinds of motion:

1. *Translational motion*, when all particles move with the same velocity and describe parallel trajectories, so that the body stays parallel to itself at all times.
2. *Rotational motion*, when the particles describe circular paths about an axis of rotation. This axis may or may not be fixed in space during the rotation of the body.

The most general motion of a rigid body is a combination of translation and rotation (notice, for example, the motion of a ball or a car wheel). An example of such a composite motion is a translation of the center of mass C of the body, with a simultaneous rotation of the body about an axis passing through C. As we will see, the center of mass plays an important role in the dynamics of a rigid body.

---

[1] More generally, a rigid body may consist of mobile parts. We will examine this case in Sect. 7.7.

C. J. Papachristou, *Introduction to Mechanics of Particles and Systems*, https://doi.org/10.1007/978-3-030-54271-9_7

7.2   Center of Mass of a Rigid Body

We have seen (Sect. 6.2) that the center of mass C of a system of particles moves in space as if it were a particle of mass equal to the total mass M of the system, subject to the total external force acting on the system. The same is true for a rigid body if this body is viewed as a structure composed of a number of particles of masses m_i. Let us assume that the only external forces acting on the system (or the rigid body) are those due to gravity. The total external force is then equal to the *total weight* of the system:

$$\vec{w} = \sum_i \vec{w}_i = \sum_i (m_i \vec{g}) = \left(\sum_i m_i \right) \vec{g} \quad \Rightarrow$$

$$\vec{w} = M \vec{g} \quad \text{where} \quad M = \sum_i m_i \tag{7.1}$$

The acceleration of gravity, $\vec{g}$, is constant in a limited region of space where the gravitational field may be considered uniform.

Note that $\vec{w}$ is a sum of forces acting on separate particles m_i located at various points of space. The question now is whether there exists some specific point of application of the total weight $\vec{w}$ of the system and, in particular, of a rigid body. A reasonable assumption is that this point could be the center of mass C of the body, given that, as mentioned above, the point C behaves as if it concentrates the entire mass M of the body and the total external force acting on it. And, in our case, $\vec{w}$ is indeed the total external force, due to gravity alone.

There is a subtle point here, however. In contrast to a point particle (such as the hypothetical "particle" of mass M moving with the center of mass C) that simply changes its location in space, a rigid body may execute a more complex motion; specifically, a combination of translation and rotation. The *translational* motion of the body under the action of gravity is indeed represented by the motion of the center of mass C, if this point is regarded as a "particle" of mass M on which the total force $\vec{w}$ is applied. For the *rotational* motion of the body, however, it is the *torques* of the external forces, rather than the forces themselves, that are responsible. Where should we place the total force $\vec{w}$ in order that the rotational motion it produces on the body be the same as that caused by the simultaneous action of the elementary gravitational forces $\vec{w}_i = m_i \vec{g}$? Equivalently, where should we place $\vec{w}$ in order that its torque *with respect to any point O* be equal to the total torque of the $\vec{w}_i$ with respect to O?

You may have guessed the answer already: at the center of mass C (this will be shown analytically in Appendix A). In conclusion:

> By placing the total weight $\vec{w}$ of the body at the center of mass C, we manage to describe both the translational and rotational motion of the body under the action of gravity.

It is for this reason that C is frequently called the *center of gravity* of the body. Note that this point does *not necessarily* belong to the body (consider, for example, the cases of a ring or a spherical shell).

In most cases a rigid body is an object exhibiting *continuous* mass distribution. Such an object can be considered as a system consisting of an enormous (practically infinite) number of particles of infinitesimal masses dm_i, placed in such a way that the distance between any two neighboring particles is zero. The total mass of the body is

$$M = \sum_i dm_i = \int dm \qquad (7.2)$$

where the sum has been replaced by an integral due to the fact that the dm_i are infinitesimal and the distribution of mass is continuous.

A point in a rigid body can be specified by its position vector $\vec{r}$, or its coordinates (x, y, z), relative to the origin O of some frame of reference. Let dV be an infinitesimal volume centered at $\vec{r} \equiv (x, y, z)$, and let dm be the infinitesimal mass contained in this volume element. The *density* ρ of the body at point $\vec{r}$ is defined as

$$\rho(\vec{r}) = \rho(x, y, z) = \frac{dm}{dV} \qquad (7.3)$$

Then,

$$dm = \rho(\vec{r})\, dV$$

and therefore, relation (7.2) for the total mass of the body is written:

$$M = \int \rho(\vec{r})\, dV \qquad (7.4)$$

where the integration takes place over the entire volume of the body (The integral is in fact a *triple* one, since, in Cartesian coordinates, $dV = dx\, dy\, dz$.). The center of mass C of the body is found by using (6.2):

$$\vec{r}_C = \frac{1}{M} \sum_i (dm_i)\, \vec{r}_i = \frac{1}{M} \int \vec{r}\, dm \quad \Rightarrow$$

$$\vec{r}_C = \frac{1}{M} \int \vec{r}\, \rho(\vec{r})\, dV \qquad (7.5)$$

where the $\vec{r}$ and $\vec{r}_C$ are measured relative to the origin O of our coordinate system (Remember, however, that the location of C *with respect to the body* is uniquely determined and is independent of the choice of the reference point O.).

In a *homogeneous* body the density has a constant value ρ, independent of $\vec{r}$. Then,

$$M = \int \rho\, dV = \rho \int dV = \rho V \qquad (7.6)$$

where V is the total volume of the body. Also, from (7.5) we have:

$$\vec{r}_C = \frac{\rho}{M} \int \vec{r} \, dV = \frac{1}{V} \int \vec{r} \, dV \tag{7.7}$$

Imagine now that, instead of a mass distribution in space, we have a *linear* distribution of mass (e.g., a very thin rod) along the x-axis. We define the *linear density* of the distribution as

$$\rho(x) = \frac{dm}{dx} \tag{7.8}$$

The total mass of the distribution is

$$M = \int dm = \int \rho(x) \, dx \tag{7.9}$$

The position of the center of mass of the distribution is given by

$$x_C = \frac{1}{M} \int x \, dm = \frac{1}{M} \int x \, \rho(x) \, dx \tag{7.10}$$

If the density ρ is constant, independent of x, then

$$M = \int \rho \, dx = \rho \int dx = \rho \, l \tag{7.11}$$

where l is the total length of the distribution. Furthermore,

$$x_C = \frac{\rho}{M} \int x \, dx = \frac{1}{l} \int x \, dx \tag{7.12}$$

As an example, consider a thin, homogeneous rod of length l, placed along the x-axis from $x = a$ to $x = a + l$, as shown in Fig. 7.1. By Eq. (7.12),

$$x_C = \frac{1}{l} \int_a^{a+l} x \, dx = \frac{1}{2l} \left[(a+l)^2 - a^2 \right] = a + \frac{l}{2}.$$

That is, the center of mass C of the rod is located at the center of the rod. Notice that the location of C on the rod is uniquely determined, independently of the choice of the origin O of the x-axis (although the value of the coordinate x_C does, of course, depend on this choice).

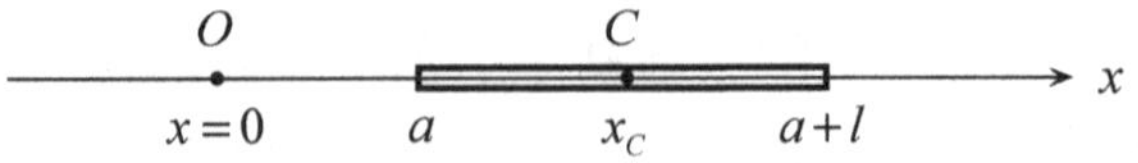

Fig. 7.1 The center of mass of a thin, homogeneous rod is at the center of the rod

7.3 Revolution of a Particle About an Axis

As we have said, a rigid body can be viewed as a system of particles m_i (or dm_i, for a continuous mass distribution) the relative positions and distances of which are fixed. So, before we study the rotational motion of a rigid body, it would be helpful to examine the revolution of a single particle m about an axis. The following analysis is fairly detailed and most of it may be skipped in a first reading; the student may thus concentrate on the main physical conclusions.

We choose the z-axis of our coordinate system to coincide with the axis of revolution and we call R the (constant) perpendicular distance of m from this axis (see Fig. 7.2). The particle m thus describes a circle of radius R, centered at some point O' of the z-axis, where O' is the normal projection of m on the axis. The axes x' and y' in the right figure are parallel to the axes x and y, respectively, while the z-axis is normal to the page and the unit vector $\hat{u}_z$ is directed *outward* (toward the reader).

In accordance with the conventions established in Sects. 2.6 and 2.7, the positive direction of traversing the circle is determined by the direction of the unit tangent vector $\hat{u}_T$ and is independent of the actual direction of motion (in Fig. 7.2 the motion is in the positive direction). By convention, the positive direction of revolution is related to the direction of $\hat{u}_z$ by means of the right-hand rule; that is, by rotating our fingers in the positive direction of revolution (i.e., in the direction of $\hat{u}_T$) our right thumb points in the direction of $\hat{u}_z$. If $\vec{R}$ is the position vector of m relative to O', and if $\hat{u}_R$ is the unit vector in the direction of $\vec{R}$ (so that $\vec{R} = R\hat{u}_R$), the ordered triad $(\hat{u}_R, \hat{u}_T, \hat{u}_z)$ forms a right-handed rectangular system of unit vectors. This means that

$$\hat{u}_R \times \hat{u}_T = \hat{u}_z , \quad \hat{u}_T \times \hat{u}_z = \hat{u}_R , \quad \hat{u}_z \times \hat{u}_R = \hat{u}_T \tag{7.13}$$

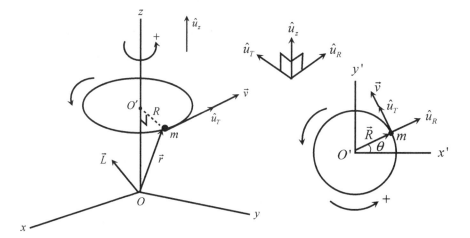

Fig. 7.2 Circular motion of a particle about the z-axis. The right figure is a view of the left figure "from above"

Fig. 7.3 The position vector
of *m*, relative to *O*, is the sum
of components in the *z*- and
R-directions

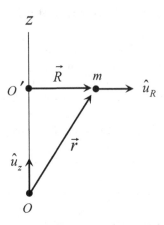

Fig. 7.3 The position vector of *m*, relative to *O*, is the sum of components in the *z*- and *R*-directions

The velocity of the particle *m* is, according to (2.32) and (2.34),

$$\vec{v} = v\,\hat{u}_T \quad \text{where} \quad v = R\,\omega = \pm|\vec{v}| \tag{7.14}$$

The angular velocity ω is positive for counterclockwise motion (as in Fig. 7.2) and negative for clockwise motion.

The position vector of *m* with respect to *O* is (see Fig. 7.3)

$$\vec{r} = \overrightarrow{OO'} + \vec{R} = z\,\hat{u}_z + R\,\hat{u}_R \tag{7.15}$$

where the distance $OO' = z$ of O' from O is one of the three coordinates of *m*. In Fig. 7.3 the unit vector $\hat{u}_T$ (which determines the positive direction of revolution) is normal to the page and directed *into* it.

The angular momentum of *m* with respect to the origin *O* of coordinates is $\vec{L} = m\,(\vec{r} \times \vec{v})$. By substituting for $\vec{r}$ and $\vec{v}$ from (7.15) and (7.14), respectively, and by using (7.13), we have:

$$\vec{L} = m\,(R\hat{u}_R + z\hat{u}_z) \times (R\omega\,\hat{u}_T) = mR^2\omega\,(\hat{u}_R \times \hat{u}_T) + mzR\omega\,(\hat{u}_z \times \hat{u}_T) \quad \Rightarrow$$
$$\vec{L} = mR^2\omega\,\hat{u}_z - mzR\omega\,\hat{u}_R \tag{7.16}$$

The *z*-component of angular momentum, i.e., the projection of $\vec{L}$ onto the *z*-axis (see Fig. 7.4) is the coefficient of $\hat{u}_z$ in (7.16):

$$L_z = mR^2\omega \tag{7.17}$$

We notice that L_z has the same sign as ω. Thus, L_z is positive or negative, depending on whether the particle revolves in the positive or the negative direction, respectively (In Fig. 7.4, L_z is positive.).

Fig. 7.4 Projection of the
angular momentum vector
onto the z-axis

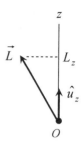

It is of interest to compare the angular momentum $\vec{L}$ relative to O with the angular
momentum $\vec{L}'$ relative to the center O' of the circular path. The position vector of m
with respect to O' is $\vec{R}$; hence, by using (7.13) and (7.14),

$$\vec{L}' = m\,(\vec{R} \times \vec{v}) = m\,(R\hat{u}_R) \times (R\omega\,\hat{u}_T) = mR^2\omega\,\hat{u}_z \qquad (7.18)$$

The z-component of $\vec{L}'$ is

$$L'_z = mR^2\omega = L_z \qquad (7.19)$$

where we have taken (7.17) into account. We notice that *the component L_z of the
angular momentum in the direction of the axis of revolution is independent of the
point of the axis relative to which the angular momentum is taken* (Of course, the
vector $\vec{L}$ of the angular momentum *does* depend on the choice of that point!). This
conclusion is of more general validity:

The component of the total angular momentum of a rotating system in the direction of the
axis of rotation assumes a unique value, independent of the choice of reference point on that
axis (i.e., of the point of the axis relative to which the angular momentum is taken).

The quantity

$$\boxed{I = mR^2} \qquad (7.20)$$

is called the *moment of inertia* of *m with respect to the axis of revolution*. Thus, (7.17)
takes on the form

$$\boxed{L_z = I\omega} \qquad (7.21)$$

Relation (7.21) connects the z-component L_z of the angular momentum with the
angular velocity ω, in the same way that the relation $p = mv$ connects the linear
momentum with the linear velocity. Note the correspondence between m and I in
these two relations: the moment of inertia I is for rotational motion what the mass
m is for linear motion.

Let $\vec{F}$ be the *total* force on m at some point of the particle's circular trajectory. The torque of $\vec{F}$ with respect to O (equal to the vector sum of the torques of all forces acting on m) is $\vec{T} = \vec{r} \times \vec{F}$. If $\vec{L}$ is the angular momentum of m relative to O, and if our reference frame, with origin O, is assumed to be inertial, then, by (3.38),

$$\vec{T} = \frac{d\vec{L}}{dt} \quad \Rightarrow$$

$$T_x\,\hat{u}_x + T_y\,\hat{u}_y + T_z\,\hat{u}_z = \frac{d}{dt}\,(L_x\,\hat{u}_x + L_y\,\hat{u}_y + L_z\,\hat{u}_z)$$

$$= \frac{dL_x}{dt}\,\hat{u}_x + \frac{dL_y}{dt}\,\hat{u}_y + \frac{dL_z}{dt}\,\hat{u}_z\;.$$

By equating the coefficients of $\hat{u}_z$ and by using (7.21), we have:

$$T_z = \frac{dL_z}{dt} = \frac{d}{dt}\,(I\omega) = I\,\frac{d\omega}{dt}$$

(since I is time-independent). Calling

$$\alpha = \frac{d\omega}{dt}$$

the angular acceleration of m, we finally have:

$$\boxed{T_z = I\alpha} \qquad\qquad (7.22)$$

If $\vec{T}' = \vec{R} \times \vec{F}$ is the torque of $\vec{F}$ with respect to O', then

$$\vec{T}' = \frac{d\vec{L}'}{dt} \quad \Rightarrow \quad T'_z = \frac{dL'_z}{dt}$$

But, by (7.19) and (7.21),

$$L'_z = L_z = I\omega$$

Thus, again,

$$T'_z = I\alpha = T_z \qquad\qquad (7.23)$$

We conclude that

the component of the total torque in the direction of the axis of revolution assumes a unique value independent of the choice of reference point on that axis.

Note the formal analogy between (7.22) and Newton's law, $F = ma$.

Relation (7.22) allows us to determine the angular acceleration of m, given the z-component of the total torque (equivalently, the torque of the total force) acting on the particle. All we need now is a practical way to evaluate T_z. First off, the particle always moves in the $x'y'$-plane, which is parallel to the xy-plane. This means that the total force $\vec{F}$ on m is a vector on the $x'y'$-plane. Such a vector can be decomposed into two orthogonal components in the directions of the unit vectors $\hat{u}_R$ and $\hat{u}_T$:

$$\vec{F} = F_R\,\hat{u}_R + F_T\,\hat{u}_T \tag{7.24}$$

In the above relation, F_T is the component of $\vec{F}$ in the tangential direction relative to the circular path of m, while F_R is the component of $\vec{F}$ in the radial direction. By using (7.15) and (7.24) we find the torque of $\vec{F}$ with respect to O:

$$\begin{aligned}
\vec{T} = \vec{r} \times \vec{F} &= (R\hat{u}_R + z\,\hat{u}_z) \times (F_R\,\hat{u}_R + F_T\,\hat{u}_T) \\
&= RF_T(\hat{u}_R \times \hat{u}_T) + zF_R(\hat{u}_z \times \hat{u}_R) + zF_T(\hat{u}_z \times \hat{u}_T) \quad \Rightarrow \\
\vec{T} &= RF_T\,\hat{u}_z + zF_R\,\hat{u}_T - zF_T\,\hat{u}_R
\end{aligned} \tag{7.25}$$

The z-component of the torque is the coefficient of $\hat{u}_z$ in (7.25):

$$\boxed{T_z = RF_T} \tag{7.26}$$

Note that T_z is independent of the location of the reference point O on the axis of revolution (since T_z is independent of z), in agreement with a remark made earlier. Assume now that $\vec{F}$ is the resultant of a set of forces $\vec{F}_1, \vec{F}_2, \ldots$, acting on m:

$$\vec{F} = \vec{F}_1 + \vec{F}_2 + \cdots = \sum_i \vec{F}_i$$

If F_{iR} and F_{iT} are the components (radial and tangential, respectively) of $\vec{F}_i$, the components F_R and F_T of the total force $\vec{F}$ are, in the spirit of Eq. (1.10),

$$F_R = \sum_i F_{iR}\,, \qquad F_T = \sum_i F_{iT}$$

By (7.26), the z-component of the total torque on m is

$$T_z = R\sum_i F_{iT} = R(F_{1T} + F_{2T} + \cdots) = RF_{1T} + RF_{2T} + \cdots$$

or

$$\boxed{T_z = T_{1z} + T_{2z} + \cdots = \sum_i T_{iz}} \tag{7.27}$$

Fig. 7.5 Revolution of a
particle subject to three
forces, about the z-axis (only
the projection O' of this axis
is shown)

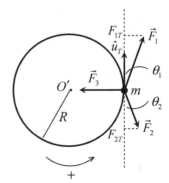

where

$$T_{iz} = RF_{iT}$$ (7.28)

is the z-component of the torque of $\vec{F}_i$. Note that (7.27) represents a sum of quantities
that may be positive or negative. Specifically, T_{iz} is positive (negative) when the
component F_{iT} of $\vec{F}_i$ in the direction of $\hat{u}_T$ is positive (negative). This, in turn, means
that $\vec{F}_i$ *tends* to cause a revolution of m in the positive (negative) direction about
the z-axis when m is initially at rest [As already mentioned, the positive direction of
revolution (i.e., the direction of $\hat{u}_T$) is related to the direction of $\hat{u}_z$ by the right-hand
rule.].

As an example, consider the case where the particle m is subject to three forces
$\vec{F}_1, \vec{F}_2, \vec{F}_3$, as shown in Fig. 7.5. In this figure the z-axis is normal to the page and
$\hat{u}_z$ is directed *outward* (toward the reader). On the other hand, $\hat{u}_T$ is, by definition,
in the positive direction of revolution, which, in turn, is related to the direction of $\hat{u}_z$
by the right-hand rule (in Fig. 7.5 the positive direction is counterclockwise).

Let F_1, F_2, F_3 be the magnitudes of the three forces. The components of these
forces in the direction of $\hat{u}_T$ (tangential components) are

$$F_{1T} = F_1 \cos\theta_1 , \quad F_{2T} = -F_2 \cos\theta_2 , \quad F_{3T} = 0$$

The signs of F_{1T} and F_{2T} are consistent with the fact that $\vec{F}_1$ tends to produce
a revolution in the positive direction, while $\vec{F}_2$ tends to generate a revolution in the
opposite direction. The vanishing of F_{3T} means that $\vec{F}_3$ (which passes through the
center O') cannot produce a revolution of m about the z-axis when m is initially at
rest. The z-components of the torques are, according to (7.28),

$$T_{1z} = RF_{1T} = RF_1 \cos\theta_1 , \quad T_{2z} = RF_{2T} = -RF_2 \cos\theta_2 , \quad T_{3z} = RF_{3T} = 0$$

The z-component of the total torque is given by (7.27):

$$T_z = T_{1z} + T_{2z} + T_{3z} = R\,(F_1\,\cos\theta_1 - F_2\,\cos\theta_2)\,.$$

Finally, the angular acceleration of m is found from (7.22) and (7.20):

$$\alpha = \frac{T_z}{I} = \frac{T_z}{mR^2} = \frac{1}{mR}\,(F_1\,\cos\theta_1 - F_2\,\cos\theta_2)$$

7.4 Angular Momentum of a Rigid Body

Consider now a rigid body rotating about an axis, which we arbitrarily choose to be the z-axis (see Fig. 7.6). As always, the direction of $\hat{u}_z$ also defines the positive direction of rotation according to the right-hand rule. We assume that the body consists of a number of particles m_i located at corresponding perpendicular distances R_i from the axis of rotation. In the course of the rotation every particle m_i executes circular motion of radius R_i, centered at the normal projection of m_i on the z-axis. *All particles have the same angular velocity* ω, equal to the angular velocity of rotation of the rigid body (Can you explain this?).

In Fig. 7.6 the body rotates in the positive direction with angular velocity ω. The angular momentum of a particle m_i, with respect to a point O of the axis of rotation, is

$$\vec{L}_i = m_i\,(\vec{r}_i \times \vec{v}_i)$$

The total angular momentum of the body, relative to O, is

Fig. 7.6 As the rigid body rotates about the z-axis, all particles composing the body revolve about this axis with common angular velocity ω

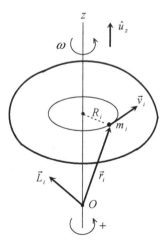

$$\vec{L} = \sum_i \vec{L}_i = \sum_i m_i (\vec{r}_i \times \vec{v}_i) \tag{7.29}$$

According to Eq. (1.10), the z-component of $\vec{L}$ equals the algebraic sum of the z-components of the $\vec{L}_i$:

$$L_z = \sum_i L_{iz}.$$

But, by (7.21) and (7.20),

$$L_{iz} = I_i \omega = m_i R_i^2 \omega.$$

Hence,

$$L_z = \sum_i (I_i \omega) = \omega \sum_i I_i$$

where we have taken into account that ω is the same for all m_i.

We now define the *moment of inertia* of the body with respect to the axis of rotation:

$$\boxed{I = \sum_i I_i = \sum_i m_i R_i^2} \tag{7.30}$$

(Notice that this is the sum of the moments of inertia of all particles making up the rigid body.) Thus, finally,

$$\boxed{L_z = I \omega} \tag{7.31}$$

Relation (7.31) gives the z-component of the angular momentum of the body, with respect to O.

It is clear from (7.30) and (7.31) that

the component of the angular momentum of a rigid body in the direction of the axis of rotation is independent of the choice of reference point on that axis, relative to which point the angular momentum is taken.

Indeed, the moment of inertia I in (7.31) is dependent only on the *perpendicular* distances R_i of the m_i from the axis of rotation, not on the distances r_i of the particles from the reference point O. For this reason we (somewhat improperly) call L_z the *"angular momentum of the rigid body with respect to the axis of rotation"*. The *vector* $\vec{L}$ of the angular momentum, of course, is only defined relative to the point O and depends, in general, on the location of O on the axis of rotation.

7.5 Rigid-Body Equations of Motion

To the extent that a rigid body can be treated as a system of particles, its motion is governed by the physical laws stated in the previous chapter. This system can perform two kinds of motion, namely, translation and rotation. The former motion is determined by the total external force acting on the body, while the latter motion is affected by the total external torque.

The momentum of a rigid body represents the total momentum of the system of particles making up the body and is given by Eq. (6.9):

$$\vec{P} = M \, \vec{v}_C \tag{7.32}$$

where M is the mass of the body and where $\vec{v}_C$ is the velocity of the body's center of mass, C. If $\vec{F}_1$, $\vec{F}_2$, ... are the external forces acting on the body (such a force is the weight $\vec{w}$, having C as its point of application) the total external force on the body is

$$\vec{F}_{\text{ext}} = \vec{F}_1 + \vec{F}_2 + \cdots = \sum \vec{F} \tag{7.33}$$

From (6.7) and (7.32) we then have:

$$\boxed{\sum \vec{F} = \frac{d\vec{P}}{dt} = M \, \frac{d\vec{v}_C}{dt} = M \, \vec{a}_C} \tag{7.34}$$

where $\vec{a}_C$ is the acceleration of the center of mass of the body. Relation (7.34) is the *equation for translational motion* of the rigid body. Note that, for translational motion it doesn't matter *where* the component forces $\vec{F}_1$, $\vec{F}_2$, ... act on the body; all we need to consider is the resultant of these forces.

On the other hand, the rotational motion of the body is determined by the total external torque. This motion, therefore, is dependent upon the points of application of the external forces. Let $\vec{T}_1$, $\vec{T}_2$, ... be the torques of $\vec{F}_1$, $\vec{F}_2$, ... relative to some point O in space. The total external torque on the body, relative to O, is

$$\vec{T}_{\text{ext}} = \vec{T}_1 + \vec{T}_2 + \cdots = \sum \vec{T} \tag{7.35}$$

Also, let $\vec{L}$ be the angular momentum of the body with respect to O, equal to the vector sum of angular momenta of all particles making up the body. According to (6.17),

$$\sum \vec{T} = \frac{d\vec{L}}{dt} \tag{7.36}$$

In Sect. 6.3 we noted that this relation is valid in either of two cases: (*a*) when O is a fixed point in some *inertial* frame of reference, or (*b*) when O coincides with the center of mass C, even if that point is accelerating (hence cannot be fixed in any inertial frame).

In the case of rotation about an axis, conventionally chosen to be the z-axis of our coordinate system, the torques and angular momenta will always be taken *with respect to a point O of that axis*. In accordance with what was said above, relation (7.36) can be used in either of two cases: (*a*) rotation about an axis passing through a point O that is either fixed or moving with constant velocity in some inertial frame, or (*b*) rotation about an axis passing through the center of mass C of the body, where C may move with constant velocity or may accelerate relative to an inertial observer, depending on whether the total external force on the body is zero or different from zero, respectively [see (7.34)].

The vector Eq. (7.36) can be resolved into three algebraic equations by taking components. We have:

$$\sum \vec{T} = \hat{u}_x \sum T_x + \hat{u}_y \sum T_y + \hat{u}_z \sum T_z$$

[see (1.10)] where, e.g., ΣT_z is the algebraic sum of the z-components of all torques acting on the body, with respect to the reference point O. Moreover,

$$\frac{d\vec{L}}{dt} = \frac{d}{dt}\,(L_x\,\hat{u}_x + L_y\,\hat{u}_y + L_z\,\hat{u}_z) = \frac{dL_x}{dt}\,\hat{u}_x + \frac{dL_y}{dt}\,\hat{u}_y + \frac{dL_z}{dt}\,\hat{u}_z$$

Equating coefficients of $\hat{u}_z$, we get:

$$\sum T_z = \frac{dL_z}{dt} \tag{7.37}$$

But, by (7.31), $L_z = I\omega$, where ω is the angular velocity of rotation and where I is the moment of inertia of the body with respect to the z-axis of rotation [see Eq. (7.30)]. By assuming that I is constant, we have:

$$\frac{dL_z}{dt} = \frac{d}{dt}\,(I\omega) = I\,\frac{d\omega}{dt} = I\alpha$$

where $\alpha = d\omega/dt$ is the angular acceleration of the rotating body. Thus, finally,

$$\boxed{\sum T_z = I\alpha} \tag{7.38}$$

Relation (7.38) is the *equation for rotational motion* of the body. Relations (7.34) and (7.38) together constitute the equations of motion of a rigid body.

From (7.38) it follows that

the component of the total external torque in the direction of the axis of rotation is independent of the reference point on that axis, relative to which point the torques of the external forces are taken.

Indeed, the moment of inertia I in (7.38) depends only on the perpendicular distances of the masses making up the body from the z-axis. For this reason the algebraic sum ΣT_z is often (albeit somewhat improperly) called the *"total torque with respect to the axis of rotation"*. Remember, however, that the *vector* representing the total external torque is always defined *with respect to a point O of the axis*.

It would be useful now to find a practical way of evaluating the sum ΣT_z. This can be accomplished by generalizing relation (7.26), as follows: Suppose that the external forces $\vec{F}_1, \vec{F}_2, \ldots$ act, respectively, at the points $P_1, P_2, \ldots$ of the body, at normal distances $R_1, R_2, \ldots$ from the z-axis of rotation.[2] As in Sect. 7.3, we call $F_{1T}, F_{2T}, \ldots$ the components of $\vec{F}_1, \vec{F}_2, \ldots$ in directions normal to the corresponding radii $R_1, R_2, \ldots$, as well as normal to the z-axis of rotation. According to (7.26), the z-components of the torques of the external forces, relative to O, are

$$T_{1z} = R_1 F_{1T}, \quad T_{2z} = R_2 F_{2T}, \ldots$$

Hence, the z-component of the total torque, equal to the sum of z-components of all external torques on the body, is

$$\sum T_z = T_{1z} + T_{2z} + \cdots = R_1 F_{1T} + R_2 F_{2T} + \cdots$$

or briefly,

$$\boxed{\sum T_z = \sum (R F_T)} \tag{7.39}$$

Note that (7.39) represents a sum of quantities that may be positive or negative. Specifically, a component F_{iT} is positive (negative) if the corresponding force $\vec{F}_i$ *tends* to cause rotation in the positive (negative) direction when the body is initially at rest. Let us see an example.

In Fig. 7.7 the z-axis of rotation is normal to the page. The unit vector $\hat{u}_z$ is directed *outward* (toward the reader), thus the positive direction of rotation is counterclockwise. The external forces $\vec{F}_1$ and $\vec{F}_2$, acting at the points P_1 and P_2 of the body, are normal to the z-axis (only the projection O of which axis is shown in the figure) and may belong to different planes parallel to each other and normal to the z-axis. The points P_1 and P_2 describe circular paths of radii R_1 and R_2, respectively, where R_1 and R_2 are the perpendicular distances of these points from the axis of rotation (The direction of rotation of the body is not specified in the figure and is of no interest to us in this particular problem.). The unit vectors $\hat{u}'_T$ and $\hat{u}''_T$, tangent to the circular paths of P_1 and P_2, respectively, always point toward the positive direction of rotation, regardless of the actual direction of rotation of the body. The tangential components of $\vec{F}_1$ and $\vec{F}_2$ are

$$F_{1T} = F_1 \cos \theta_1, \quad F_{2T} = -F_2 \cos \theta_2$$

---

[2]To simplify our analysis we assume that all external forces are perpendicular to the axis of rotation.

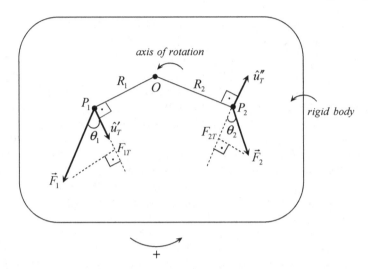

Fig. 7.7 Rotation of a body subject to two forces, about the z-axis (only the projection O of this axis is shown)

where F_1 and F_2 are the magnitudes of the two forces. The physical significance of the signs of F_{1T} and F_{2T} is that $\vec{F}_1$ tends to rotate the body in the positive direction, while $\vec{F}_2$ tends to generate a rotation in the negative direction. The z-components of the external torques are

$$T_{1z} = R_1 F_{1T} = R_1 F_1 \cos\theta_1 , \qquad T_{2z} = R_2 F_{2T} = -R_2 F_2 \cos\theta_2.$$

Thus, the total torque with respect to the axis of rotation is

$$\sum T_z = T_{1z} + T_{2z} = R_1 F_1 \cos\theta_1 - R_2 F_2 \cos\theta_2.$$

Finally, the angular acceleration of the body is, according to (7.38),

$$\alpha = \frac{1}{I} \sum T_z = \frac{1}{I} \ (R_1 F_1 \cos\theta_1 - R_2 F_2 \cos\theta_2)$$

where I is the moment of inertia of the body with respect to the axis of rotation.

7.6 Moment of Inertia and the Parallel-Axis Theorem

In the case of a rigid body consisting of a discrete set of particles $m_1, m_2, \ldots,$ the moment of inertia with respect to an axis is

Fig. 7.8 Rotating system of
two spheres connected by a
thin, weightless rod of length
L

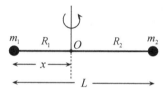

$$I = \sum_i m_i R_i^2 = m_1 R_1^2 + m_2 R_2^2 + \cdots \qquad (7.40)$$

where R_i is the perpendicular distance of m_i from the axis. As an example, consider
two spheres of masses m_1 and m_2 connected to each other by a thin, weightless rod
of length L, as seen in Fig. 7.8 (the role of the rod is simply to keep the spheres at a
constant distance L from each other; the rod is thus not counted as part of the rigid
body).

The moment of inertia of this system, with respect to an axis passing through O,
is

$$I = m_1 R_1^2 + m_2 R_2^2 = m_1 x^2 + m_2 (L - x)^2$$

where we have put $x = R_1$. In particular, if the axis of rotation passes through m_1, then
$x = 0$ and $I = m_2 L^2$. As can be proven (see below) the moment of inertia assumes a
minimum value when the axis of rotation passes through the center of mass C of the
system. It is thus easier to rotate the system about such an axis, since a given external
torque will then produce a maximum angular acceleration, according to (7.38).

For rigid bodies consisting of a continuous distribution of matter, the sum in (7.40)
must be replaced by an integral:

$$I = \int R^2 dm \qquad (7.41)$$

where R is the perpendicular distance of the elementary mass dm from the axis of
rotation. This mass is written $dm = \rho dV$, where ρ is the density of the body at
the point where dm is located, and where dV is the volume occupied by dm. For a
homogeneous body the density ρ is constant, equal to $\rho = M/V$, where M is the mass
and V is the total volume of the body. Relation (7.41) is then written:

$$I = \int R^2 \rho \, dV = \rho \int R^2 dV = \frac{M}{V} \int R^2 dV \qquad (7.42)$$

To evaluate the above integrals one must know the specific geometrical characteristics of the rigid body. A table of moments of inertia for the most common
geometries is given in Appendix C.

The *parallel-axis theorem* (or *Steiner's theorem*) allows us to calculate the moment
of inertia of a rigid body with respect to an axis, given the moment of inertia with

Fig. 7.9 A thin rod of length l, lying along the x-axis

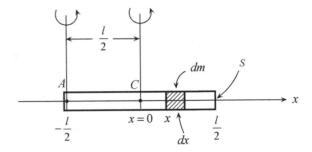

respect to a parallel axis passing through the center of mass of the body. We state this theorem without proof (see [1, 2]). Consider a rigid body of mass M. Let I be the moment of inertia of the body with respect to an axis, and let I_C be the moment of inertia with respect to a parallel axis passing through the center of mass C of the body. If the perpendicular distance between the two axes is equal to a, then

$$\boxed{I = I_C + M a^2}$$ (7.43)

According to (7.43), given an infinite set of axes parallel to one another, the moment of inertia of a rigid body becomes a *minimum* when taken with respect to the axis passing through the center of mass of the body. It is for this reason that, as mentioned earlier, it is easiest to rotate a body about an axis passing through its center of mass C (Of course, an infinite number of axes pass through C, each axis defining a separate infinity of parallel axes.).

As an application of (7.42), we will calculate the moment of inertia of a thin rod of length l and mass M (see Fig. 7.9) with respect to an axis perpendicular to the rod and passing through its center C (which point coincides with the center of mass of the rod, as shown in an example at the end of Sect. 7.2). We call S the cross-sectional area of the rod and we assume that the rod lies on the x-axis.

We assume that the center C of the rod is located at the point $x = 0$ of the axis, so that the rod extends from $x = -l/2$ to $x = l/2$ (this choice has no effect on the value of the moment of inertia, since this value depends only on the position of the axis of rotation *relative to the body*). We consider an elementary section of the rod, from x to $x + dx$. The volume of this section is $dV = Sdx$, while the total volume of the rod is $V = S l$. The distance R of the point x from the center of mass C (thus also the perpendicular distance of x from the axis that passes through C) is $R = |x|$. Relation (7.42) is written (with $I = I_C$ for an axis passing through C):

$$I_C = \frac{M}{V} S \int_{-l/2}^{l/2} x^2 dx = \frac{M}{l} \left[\frac{x^3}{3} \right]_{-l/2}^{l/2} \quad \Rightarrow$$

$$I_C = \frac{1}{12} M l^2$$ (7.44)

To find the moment of inertia of the rod with respect to a parallel axis passing through an end of it—say, the end A of the rod—we use (7.43) with $a = l/2$:

$$I_A = I_C + M a^2 = \frac{1}{12} M l^2 + \frac{1}{4} M l^2 \quad \Rightarrow$$

$$I_A = \frac{1}{3} M l^2 \tag{7.45}$$

Of course, the same result can be found directly from (7.42) (show this).

7.7 Conservation of Angular Momentum

Up to this point we have treated the moment of inertia I as a constant quantity, unchanged with time. This is true as long as the shape of the rigid body doesn't change in the course of the body's motion. We will now relax this requirement and consider rigid bodies whose shape *may* change while they move. This is, e.g., the case with the human body and, in general, with rigid bodies consisting of mobile parts. In such cases the moment of inertia with respect to an axis is a variable quantity and its change with time may significantly affect the rotational motion of the body even in the absence of external torques.

The treatment of such complex problems, as well as of many others of a different nature (e.g., problems combining collision with rotational motion; see Problems 44 and 45) is simplified by using conservation of angular momentum, provided, of course, that the conditions for validity of this principle are fulfilled.

As we know (Sect. 7.5) the angular momentum of a rigid body and the total external torque on the body are related by

$$\sum \vec{T} = \frac{d\vec{L}}{dt} \tag{7.46}$$

As stressed in Sect. 7.5, the above relation is valid for rotation about an axis passing through the common reference point O of $\vec{L}$ and $\vec{T}$, where O may be a fixed point in some inertial reference frame or may coincide with the center of mass C of the body (even if C accelerates relative to an inertial observer). We notice that, if $\sum \vec{T} = 0$, then $d\vec{L}/dt = 0 \Rightarrow \vec{L} = const.$; that is,

when the total external torque on a rigid body, relative to a point O, is zero, the angular momentum of the body relative to O is constant in time.

This conclusion constitutes the *principle of conservation of angular momentum* for a rigid body.

Note carefully that the vector relation (7.46) is always understood to be valid *with respect to a point O*. By taking the z-component of (7.46), however, we find an algebraic relation that is valid *with respect to the z-axis of rotation*, regardless of the position of the reference point O on that axis (see Sect. 7.5):

$$\sum T_z = \frac{dL_z}{dt} \tag{7.47}$$

In particular, when $\Sigma T_z = 0$, the component L_z of angular momentum is constant:

When the total external torque on the body relative to the axis of rotation is zero, the angular
momentum of the body relative to this axis is constant in time.

The above statement expresses the *principle of conservation of angular momentum
with respect to the axis of rotation*.

We remark that the choice of the z-axis of our coordinate system as coincident
with the axis of rotation is totally arbitrary! We could have named our axes differently
so that, e.g., the axis of rotation would be the x-axis or the y-axis. Relation (7.47)
would then have to be rewritten with x or y in place of z.

Now, according to (7.31), $L_z = I\omega$, where ω is the angular velocity of rotation and
where I is the moment of inertia of the rigid body relative to the z-axis of rotation.
Thus, the constancy of angular momentum with respect to this axis is expressed as

$$I\omega = \text{constant} \quad \Leftrightarrow \quad I_1\omega_1 = I_2\omega_2 \tag{7.48}$$

where the indices 1 and 2 refer to two moments t_1 and t_2. If I is constant ($I_1 = I_2$),
which occurs when the geometry of the body is unchanged, the angular velocity ω is
constant in time ($\omega_1 = \omega_2$). This conclusion also follows from (7.38), according to
which the angular acceleration α is zero (hence, the angular velocity ω is constant)
when the total torque with respect to the axis of rotation vanishes.

We talked earlier about a basic difference between the vector relation (7.46) and
the algebraic relation (7.47). In (7.46) the $\vec{L}$ and $\vec{T}$ are evaluated with respect to a
point O of the axis of rotation and, generally, depend on the choice of this point,
whereas in (7.47) the L_z and T_z are taken with respect to the *axis* itself and do *not*
depend on the location of the reference point O on this axis. For each body, however,
there is always a special set of axes of rotation, each of which passes through the
center of mass C of the body and possesses the following property: the vector $\vec{L}$ of the
angular momentum of the body [thus, by (7.46), the total external torque $\Sigma \vec{T}$ also]
does not depend on the choice of reference point O on that axis but assumes a unique
value for *all* points of the axis. Furthermore, the angular momentum $\vec{L}$ is directed
parallel to the axis. An axis of rotation having these properties is called a *principal
axis*. In particular, every axis of symmetry passing through the center of mass C of
the body is a principal axis for this body. Thus, in the case of a sphere, every axis
passing through its center is a principal axis. For a cylinder, the central axis as well
as every axis normal to it and going through the center of mass are principal axes.
For a cube, the three axes normal to the faces and passing through the center of the
cube are principal axes (More on principal axes will be said in Appendix D. See also
[2, 3].).

Assume now that the z-axis of rotation is a principal axis of the rigid body (as mentioned above, such an axis always passes through the center of mass C of the body). The angular momentum $\vec{L}$ of the body, with respect to *any* point of this axis, will be independent of the location of that point on the axis and equal to

$$\vec{L} = I\,\omega\,\hat{u}_z \tag{7.49}$$

where ω is the angular velocity of rotation and where I is the moment of inertia with respect to the principal axis (remember that ω may be positive or negative, depending on the direction of rotation).

We define the *angular velocity vector*

$$\vec{\omega} = \omega\,\hat{u}_z \tag{7.50}$$

Equation (7.49) is then written:

$$\boxed{\vec{L} = I\,\vec{\omega}} \tag{7.51}$$

Relation (7.51) gives the angular momentum of a rigid body with respect to *any* point of a *principal* axis. Note that the angular momentum is directed parallel to the principal axis.

It must be noted here that a relation of the form (7.51) may sometimes be valid for axes that are *not* principal. For example, (7.51) gives the angular momentum $\vec{L}$ of a thin flat plate rotating about an axis perpendicular to the plate, or of a rod rotating about an axis normal to it. Careful, however: In each of these examples the angular momentum $\vec{L}$ is taken *with respect to the point O of the body through which the axis passes*! For any other point, (7.51) is *not* valid, unless the axis of rotation is a *principal* axis (such as is, e.g., the axis normal to a circular disk and passing through the center of the disk, or, the axis intersecting a rod perpendicularly at its center).

Let us now assume that, for any of the above reasons, relation (7.51) is valid relative to a point O of the rotation axis. The fundamental Eq. (7.46) is then written, relative to this point,

$$\sum \vec{T} = \frac{d\vec{L}}{dt} = \frac{d}{dt}\,(I\,\vec{\omega}) \tag{7.52}$$

If the moment of inertia I is constant,

$$\boxed{\sum \vec{T} = I\,\frac{d\vec{\omega}}{dt} = I\,\vec{\alpha}} \tag{7.53}$$

where $\vec{\alpha}$ is the *angular acceleration vector*. If it happens that the total external torque on the body, relative to O, is zero ($\Sigma\,\vec{T} = 0$), then, by (7.52), the angular momentum $\vec{L}$ with respect to that point is constant:

$$\vec{L} = I\,\vec{\omega} = \text{constant} \quad \Leftrightarrow \quad I_1\vec{\omega}_1 = I_2\vec{\omega}_2 \tag{7.54}$$

Moreover, if I is constant, the angular acceleration $\vec{\alpha}$ is zero and the angular velocity $\vec{\omega}$ is constant, as follows from (7.53) and (7.54).

Let us see some examples:

1. Consider a body of mass M moving in space under the sole action of gravity; that is, no forces act on the body other than its weight $\vec{w} = M\,\vec{g}$. Hence, as explained in Sect. 7.2, the total external force on the body can be considered acting at the center of mass C (even if that point does not belong to the body, as, e.g., in the cases of a ring or a spherical shell). Then, the total torque with respect to C is zero and, according to (7.46), the angular momentum $\vec{L}$ of the body relative to C is constant:

> If a body moves under the sole action of gravity, the angular momentum of the body relative to its center of mass is constant.

What does a diver do in order to increase his angular speed and make several somersaults in the air? He allows his body to contract as much as possible by pulling hands and feet close to the trunk of his body, thus decreasing his moment of inertia relative to the axis of rotation (which axis is horizontal and passes through the center of mass C of the athlete). According to (7.48) and (7.54), this results in an increase of the magnitude of the angular velocity of rotation, $\vec{\omega}$, of the body [We assume that (7.51) is approximately valid when $\vec{L}$ is taken with respect to C.].

2. You may have observed the spinning motion of a figure skater on ice. The skater is subject to two forces, namely, her weight and the normal reaction from the ice (we assume that friction is negligible). None of these forces produces torque with respect to the center of mass C of the skater; hence, the angular momentum of the skater relative to C is constant. Moreover, the vertical axis of rotation is a principal axis, relative to which the conservation of angular momentum is expressed in the form (7.54) [or, algebraically, in the form (7.48)]. To increase her angular speed, the skater pulls her hands close to the trunk of her body, which has the effect of decreasing her moment of inertia. She does exactly the opposite in order to decrease her angular speed.

3. Why is a bicycle harder to overturn when it is in motion? Consider, for simplicity, a single bicycle wheel and call $\vec{L}$ the angular momentum of the wheel with respect to its (principal) axis of rotation at time t. If $\vec{T}$ is the total external torque on the wheel at this instant, the change of angular momentum within an infinitesimal time interval dt is $d\vec{L} = \vec{T}\,dt$. Now, an overturn of the wheel is accompanied by a change of direction of $\vec{L}$, which change requires a torque $\vec{T}$ *perpendicular* to $\vec{L}$ (in the same way that a force perpendicular to the velocity of a body is needed to produce a change of the direction of motion of the body). Then, the infinitesimal change $d\vec{L}$ of the angular momentum will be normal to $\vec{L}$, as shown in Fig. 7.10.[3]

We notice that

---

[3] The figure is not drawn to scale. In fact, if the torque is normal to the angular momentum, then for $dt \to 0$ the magnitudes of the latter at times t and $t + dt$ tend to be equal (see Sect. 7.12).

Fig. 7.10 A torque normal
to the angular momentum is
needed in order to change the
direction of the latter

$$\tan\theta = \frac{|d\vec{L}|}{|\vec{L}|} = \frac{|\vec{T}|\,dt}{|\vec{L}|} \tag{7.55}$$

Therefore, the larger the magnitude of the angular momentum $\vec{L}$ of the wheel [thus, the higher the angular speed ω of the wheel, according to (7.51)] the smaller will be the angle of deflection θ for a given torque $\vec{T}$, and the more stable will be the motion of the wheel. It is for this reason that, the faster a bicycle moves, the harder it is to overturn.

7.8 Equilibrium of a Rigid Body

A body is in *translational equilibrium* if the total external force on it is zero. The body is in *rotational equilibrium* if the total external torque on it, relative to *any* point of space, is zero. We write:

$$\sum \vec{F} = 0 \quad (a)$$
$$\sum \vec{T} = 0 \quad (b) \tag{7.56}$$

Think twice before you ask yourself the (supposedly rhetorical) question, *"what need do we have of condition (b) if condition (a) is already satisfied?"*! A vanishing total force does *not* necessarily imply a vanishing total torque, and vice versa. Here are two examples:

1. Assume that a body is subject to two forces of equal magnitudes but opposite directions, acting along parallel lines (Fig. 7.11). Such a system of forces $\vec{F}_1$ and $\vec{F}_2$, where $\vec{F}_2 = -\vec{F}_1$, is called a *couple*.

The total force on the body is

Fig. 7.11 Two opposite
forces forming a couple

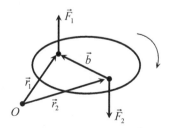

$$\sum \vec{F} = \vec{F}_1 + \vec{F}_2 = \vec{F}_1 + (-\vec{F}_1) = 0.$$

According to (7.34), the velocity of the center of mass C of the body is constant:

$$\sum \vec{F} = M \vec{a}_C = M \frac{d\vec{v}_C}{dt} = 0 \quad \Rightarrow \quad \vec{v}_C = \text{constant}.$$

Hence, if C is initially at rest ($\vec{v}_C = 0$) it will remain at rest. We say that the body is in translational equilibrium.

On the other hand, the total torque with respect to an arbitrary point O (Fig. 7.11) is

$$\sum \vec{T} = \vec{T}_1 + \vec{T}_2 = \vec{r}_1 \times \vec{F}_1 + \vec{r}_2 \times \vec{F}_2 = \vec{r}_1 \times \vec{F}_1 - \vec{r}_2 \times \vec{F}_1 = (\vec{r}_1 - \vec{r}_2) \times \vec{F}_1.$$

Putting $\vec{b} = \vec{r}_1 - \vec{r}_2$, we have:

$$\sum \vec{T} = \vec{b} \times \vec{F}_1 \qquad (7.57)$$

Relation (7.57) gives the *torque of a couple*. Obviously, this torque is *independent of the choice of reference point O*. We note that the total torque on the body is different from zero, even though the resultant force *is* zero! This torque will produce a rotation of the body about its center of mass C, which point, as argued above, will remain at rest if initially at rest. If the axis of rotation is a principal axis, we can find the angular acceleration $\vec{\alpha}$ of the body by using (7.53) and (7.57):

$$\vec{\alpha} = \frac{1}{I} \sum \vec{T} = \frac{1}{I} (\vec{b} \times \vec{F}_1) \qquad (7.58)$$

2. Consider now a body of constant shape, subject only to its weight $\vec{w}$. Thus,

$$\sum \vec{F} = \vec{w} \neq 0.$$

The center of mass C of the body moves relative to an inertial observer with acceleration equal to

$$\vec{a}_C = \frac{1}{M} \sum \vec{F} = \frac{\vec{w}}{M} = \vec{g}.$$

On the other hand, the single force $\vec{w}$ passes through C, and so the total torque on the body, relative to C, is zero ($\sum \vec{T} = 0$). Consequently, the body does not acquire angular acceleration with respect to its center of mass. We say that the body is in rotational equilibrium about C.

We thus conclude that an absolute state of equilibrium of a body requires that *both* relations (7.56) be satisfied simultaneously. These vector relations are equivalent to a system of six algebraic equations:

$$\sum F_x = 0 \,, \qquad \sum F_y = 0 \,, \qquad \sum F_z = 0 \qquad (7.59)$$

$$\sum T_x = 0 \,, \qquad \sum T_y = 0 \,, \qquad \sum T_z = 0 \qquad (7.60)$$

where F_x, T_x, etc., are the components of the various forces and torques acting on the body. In particular, relations (7.60) must be satisfied independently of the choice of the origin O of the coordinate system (x, y, z), i.e., must be valid with respect to *any* point of reference O of the torques. A body subject to the above conditions will move with constant momentum and constant velocity of its center of mass, as well as with constant angular momentum relative to *any* point of space.

7.9 Kinetic and Total Mechanical Energy

Consider a rigid body rotating with angular velocity ω about an axis passing through a fixed point O of space (Fig. 7.12). During rotation, every elementary mass m_i in the body moves circularly about the axis of rotation with the common angular velocity ω. If R_i is the perpendicular distance of m_i from the axis (thus, the radius of the circular path of m_i) the speed of this mass element is $v_i = R_i \, \omega$. The total *kinetic energy of rotation* is the sum of the kinetic energies of all elementary masses m_i contained in the body:

$$E_{k,\,\mathrm{rot}} = \sum_i \left(\frac{1}{2} m_i v_i^2 \right) = \sum_i \left(\frac{1}{2} m_i R_i^2 \omega^2 \right) = \frac{1}{2} \omega^2 \sum_i m_i R_i^2 \quad \Rightarrow$$

Fig. 7.12 All elementary masses in the body revolve about the axis of rotation with common angular velocity ω

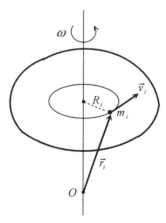

$$\boxed{E_{k,\,\mathrm{rot}} = \frac{1}{2} I \,\omega^2}$$

(7.61)

where

$$I = \sum_i m_i R_i^2$$

is the moment of inertia of the body relative to the axis of rotation.

Relation (7.61) represents the total kinetic energy of a body when this body performs *pure rotation* about a fixed axis. A more general kind of motion is a rotation about an axis that is moving in space. Specifically, assume that the axis of rotation passes through the center of mass C of the body, while C itself moves in space with velocity $\vec{v}_C$. The body thus performs a composite motion consisting of a *translation* of the center of mass C and a *rotation* about C. According to a remark made at the end of Sect. 6.4, the total kinetic energy of the body is the sum of two quantities: a *kinetic energy of translation*,

$$E_{k,\,\mathrm{trans}} = \frac{1}{2} M v_C^2$$

(7.62)

(where M is the mass of the body and v_C is the speed of the center of mass C) and a *kinetic energy of rotation about C*,

$$E_{k,\,\mathrm{rot}} = \frac{1}{2} I_C \,\omega^2$$

(7.63)

(where ω is the angular velocity of rotation about an axis passing through C, while I_C is the moment of inertia of the body relative to that axis). The total kinetic energy of the body is, therefore,

$$\boxed{E_k = E_{k,\,\mathrm{trans}} + E_{k,\,\mathrm{rot}} = \frac{1}{2} M v_C^2 + \frac{1}{2} I_C \,\omega^2}$$

(7.64)

If the body is subject to external forces that are conservative, we can define an *external potential energy* E_p as well as a *total mechanical energy* E, the latter assuming a constant value during the motion of the body:

$$\boxed{E = E_k + E_p = \frac{1}{2} M v_C^2 + \frac{1}{2} I_C \,\omega^2 + E_p = const.}$$

(7.65)

For example, if the body moves under the sole action of gravity, its potential energy is

$$E_p = M g \, y_C$$

(7.66)

where y_C is the vertical distance (the height) of the center of mass C with respect to an arbitrary horizontal plane of reference. Indeed, by relation (6.3),

$$y_C = \frac{1}{M} \sum_i m_i y_i$$

where y_i is the height above the reference plane, of the location of the elementary mass m_i in the body. The total gravitational potential energy of the body, equal to the sum of the potential energies of all elementary masses m_i, is then

$$E_p = \sum_i (m_i g\, y_i) = g \sum_i m_i y_i = M g\, y_C.$$

The total mechanical energy of the body is constant and equal to

$$E = \frac{1}{2} M v_C^2 + \frac{1}{2} I_C \omega^2 + M g\, y_C \qquad (7.67)$$

7.10 Rolling Bodies

The rolling of a body (such as a sphere or a cylinder) on a plane surface can be described in two equivalent ways:

a. *Combination of translation and rotation*

Rolling can be considered as a composite motion consisting of a rotation about an axis parallel to the surface and passing through the center of mass C of the body, with a simultaneous translation of this axis parallel to itself and in such a way that the axis is always normal to the (constant) direction of motion of C. Figure 7.13 shows a cross-section of the body, passing through C and normal to the surface and to the axis of rotation. The axis is normal to the page, so that only its projection C is visible in the figure. We call ω the angular velocity and $\alpha = d\omega/dt$ the angular acceleration of the rolling body relative to the axis of rotation.

Fig. 7.13 Cross-section of a rolling body (The plane of rolling is not necessarily horizontal!)

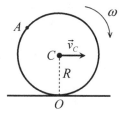

Let A be an arbitrary point of the circumference of the cross-section (obviously, A belongs to the surface of the body). The velocity $\vec{v}_C$ of the center of mass with respect to the plane surface on which rolling takes place (*plane of rolling*) is normal to the axis of rotation; so, in Fig. 7.13 the vector $\vec{v}_C$ belongs to the cross-section passing through C and containing the point A.

We denote by O the point of the circumference of the cross-section that is *momentarily* in contact with the plane of rolling. Equivalently, O can be regarded as a point of that plane. Also, we denote the velocity and the acceleration of C with respect to the plane (or, if you prefer, with respect to the point of contact O) by

$$\vec{v}_{C,O} \equiv \vec{v}_C, \qquad \vec{a}_{C,O} \equiv \vec{a}_C.$$

The magnitudes of the velocity and the *tangential* acceleration of A, relative to the center of mass C, are

$$v_{A,C} = R\omega, \qquad a_{A,C} = R\alpha \tag{7.68}$$

where R is the radius of the cross-section, and where we have used (2.34) and (2.36).

The motion is called *rolling without slipping*—or, simply, *pure rolling*—if the body does not slide on the plane of rolling. This means that the point of the body in contact with the plane of rolling does not move along the plane but its contact with this plane is only instantaneous. The *condition for pure rolling* can be expressed as follows:

The velocity and the tangential acceleration of a point A of the circumference of the cross-section, relative to the center of mass C, are equal in magnitude to the velocity and the acceleration, respectively, of C with respect to the plane of rolling (or, with respect to the instantaneous point of contact O).

We write:

$$v_{A,C} = v_{C,O} \equiv v_C \ , \qquad a_{A,C} = a_{C,O} \equiv a_C \tag{7.69}$$

By combining (7.68) and (7.69), the condition for pure rolling is written:

$$v_C = R\omega \ , \qquad a_C = R\alpha \tag{7.70}$$

We can justify this condition as follows: In rolling without slipping, as the point A of the circumference of the cross-section describes an arc of length s relative to the center C, the point C itself travels the same distance s relative to the plane of rolling. For example, when the body performs a complete turn about C, the point A describes an arc of length $s = 2\pi R$ with respect to C. The same distance is traveled, in the meanwhile, by C with respect to the plane (Observe, e.g., the motion of a car wheel that doesn't slip on the road.). Therefore, the velocity of A relative to C must be equal in magnitude to the velocity of C relative to the point of contact O; that

Fig. 7.14 Pure rolling
viewed as a rotation about an
instantaneous axis passing
through the point of contact
O (only the projection O of
this axis is visible)

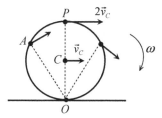

is, $v_C = R\omega$. By differentiating this relation with respect to time, and by taking into
account that $dv_C/dt = a_C$ and $d\omega/dt = \alpha$, we get: $a_C = R\,\alpha$.

b. *Rotation about an instantaneous axis*

Alternatively, pure rolling may be viewed as a rotation about an *instantaneous* axis
belonging to the plane of rolling, passing through the point of contact O and normal
to the velocity of the center of mass C (see Fig. 7.14).

At any moment, every point A of the cross-section of the body tends to move
on a circle with center the instantaneous point of contact O, of radius equal to the
distance OA of the considered point from O (note that A may now be *any* point of
the cross-section, not necessarily a point of the circumference). We notice that the
angular velocity of A with respect to O is the same as the angular velocity of A with
respect to C; that is, equal to ω. Indeed, the angular velocity of A relative to O is the
same as the angular velocity of any other point of the cross-section relative to O; in
particular, of the center of mass C. But, the angular velocity of C with respect to O
is equal to the angular velocity of O with respect to C, which, in turn, is equal to ω
(Think of it as follows: Within an infinitesimal time interval dt, the angle described
by C relative to the point O of the plane of rolling is the same as the angle described
by the point O of the body relative to C.).

Given that, momentarily, the motion of A is circular about O, the speed of A
relative to O (that is, relative to the plane of rolling) is

$$v_{A,O} = (OA)\,\omega \qquad\qquad (7.71)$$

while the direction of motion of A is normal to the radius OA. If A coincides with O,
then $(OA) = 0$ and

$$v_{O,O} = 0 \qquad\qquad (7.72)$$

On the other hand, if A coincides with C, then $(OA) = (OC) = R$ and

$$v_{C,O} \equiv v_C = R\omega \qquad\qquad (7.73)$$

By differentiating this with respect to time, we get:

$$a_{C,O} \equiv a_C = R\alpha \tag{7.74}$$

We have thus recovered the condition (7.70) for pure rolling. Finally, if A coincides with the top point P of the cross-section (see Fig. 7.14) then $(OA) = (OP) = 2R$ and

$$v_{P,O} \equiv v_P = 2R\omega = 2v_C \tag{7.75}$$

Differentiating this with respect to time, we have:

$$a_{P,O} \equiv a_P = 2R\alpha = 2a_C \tag{7.76}$$

7.11 The Role of Static Friction in Rolling

In many cases (though not always) friction is necessary for rolling on a surface. Imagine, for example, a car at rest attempting to start moving on an icy road! The role of friction can be appreciated with the aid of the following example.

Consider a cylinder that is rolling without slipping on an inclined plane (Fig. 7.15). The angular velocity of the cylinder about the axis of rotation (which is a principal axis passing through the center of mass C) increases as the cylinder rolls down the incline, which indicates the presence of an external torque with respect to C. What force can be responsible for that torque? Certainly, neither the weight $\vec{w}$ nor the normal reaction $\vec{N}$ from the plane, since both these forces pass through C. The only remaining force is *static friction* $\vec{f}$, the role of which is to prevent the cylinder from sliding on the incline (otherwise the motion would not be a pure rolling and, in that case, friction would be *kinetic*). We thus conclude that pure rolling would be impossible on an inclined plane without the presence of static friction (This conclusion is not valid, however, if the plane of rolling is horizontal; see Problem 47.).

It should be noted that

in pure rolling the static friction does not produce work and thus has no effect on the conservation of mechanical energy.

[This is in contrast to rolling *with* slipping, where the friction is kinetic and does produce (negative) work; see Sect. 4.7.] Indeed, in pure rolling the point of application

Fig. 7.15 Static friction is needed for pure rolling of a cylinder on an inclined plane

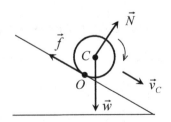

O of the static friction $\vec{f}$ does not move along the plane of rolling, since, according to (7.72), the instantaneous velocity of this point relative to the plane is zero. No work is thus produced by $\vec{f}$.

7.12 Gyroscopic Motion

Generally speaking, the rotational motion of a body is characterized as *gyroscopic* if the axis of rotation passes through a fixed point of space but the direction of the axis changes with time. If the axis of rotation is a principal axis (e.g., an axis of symmetry, passing through the center of mass of the body) then, according to (7.49), the angular momentum $\vec{L}$ of the body is directed parallel to that axis. Thus, in gyroscopic motion about a principal axis the direction of the angular momentum may change with time.

In general, a change of the body's angular momentum $\vec{L}$ requires an external torque $\vec{T}$, where the vectors $\vec{L}$ and $\vec{T}$ are evaluated with respect to any point of the principal axis of rotation and are independent of the choice of that point. According to (7.46),

$$\vec{T} = \frac{d\vec{L}}{dt} \quad \Leftrightarrow \quad d\vec{L} = \vec{T}\,dt.$$

We notice that the infinitesimal change $d\vec{L}$ of the angular momentum is in the direction of the external torque.

If the torque $\vec{T}$ is *normal* to the angular momentum $\vec{L}$ (see Fig. 7.10) the change $d\vec{L}$ is normal to $\vec{L}$, so that $\vec{L} \cdot d\vec{L} = 0$. But,

$$\vec{L} \cdot d\vec{L} = \frac{1}{2}\,d(\vec{L} \cdot \vec{L}) = \frac{1}{2}\,d(L^2) = \frac{1}{2}\,(2L\,dL) = L\,dL$$

(where L is the magnitude of $\vec{L}$) and, therefore, $L\,dL = 0$. Given that $L \neq 0$, we conclude that $dL = 0 \Leftrightarrow L = $ constant:

If the external torque is normal to the body's angular momentum, the magnitude of the angular momentum is constant in time (it is only the direction of $\vec{L}$ that changes).

This conclusion reminds us of the constancy of a particle's speed (magnitude of the velocity) when the total force on the particle is a vector perpendicular to the velocity. Here we can think of the angular momentum $\vec{L}$ of the body as a vector of constant magnitude L, having a direction that changes with time. The direction of the principal axis of rotation then also changes accordingly. It is often the case that the axis of rotation is itself precessing about another axis that is fixed in space. This kind of gyroscopic motion occurs, for example, in the case of a *spinning top* (Fig. 7.16).

In general, bodies capable of executing gyroscopic motion are called *gyroscopes*. An important application is the *gyroscopic compass*, with the aid of which one can

Fig. 7.16 A spinning top. The force of gravity acts on the center of mass C, producing a torque in the direction perpendicular to the angular momentum. The latter vector precesses about the z-axis, retaining a constant magnitude

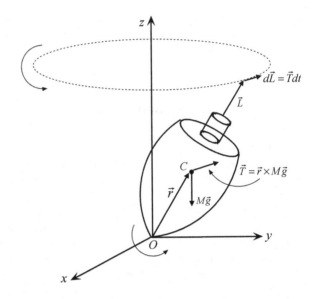

determine the direction to the North. Because of the rotation of the Earth about its axis, an external torque is exerted on the gyroscope of the compass, forcing the axis of rotation of the gyroscope to be aligned with the axis of rotation of the Earth (see, e.g., Sect. 13.10 of [4]).

Comparative Table of Translational and Rotational Motion

Translational	Rotational
$\vec{a} = \frac{d\vec{v}}{dt}$	$\vec{\alpha} = \frac{d\vec{\omega}}{dt}$
$\vec{p} = m\,\vec{v}$	$\vec{L} = I\,\vec{\omega}$
$\vec{F} = \frac{d\vec{p}}{dt} = m\,\vec{a}$	$\vec{T} = \frac{d\vec{L}}{dt} = I\,\vec{\alpha}$
$E_{k,\text{trans}} = \frac{1}{2}m\,v^2$	$E_{k,\text{rot}} = \frac{1}{2}I\,\omega^2$

References

1. K.R. Symon, *Mechanics*, 3rd edn. (Addison-Wesley, 1971)
2. J.B. Marion, S.T. Thornton, *Classical Dynamics of Particles and Systems*, 4th edn. (Saunders College, 1995)
3. J.R. Taylor, *Classical Mechanics* (University Science Books, 2005)
4. M. Alonso, E.J. Finn, *Physics* (Addison-Wesley, 1992)

Chapter 8
Elementary Fluid Mechanics

Abstract The concept of an ideal fluid is defined and various principles of Hydro-statics (such as Pascal and Archimedes' principles) are stated and proven. Basic equations of Hydrodynamics (equation of continuity, Bernoulli's equation) are presented and their physical content is examined.

8.1 Ideal Fluid

The term "*fluid*" signifies a continuous medium that can flow. Depending on their physical properties (such as, e.g., compressibility) fluids are separated into *liquids* and *gases*. In this chapter we focus our attention to the study of liquids, and it is in this sense that the term "*fluid*" will be used henceforth. The mechanics of fluids can be separated into two parts; namely, *Fluid Statics* or *Hydrostatics*, which studies fluids at rest, and *Fluid Dynamics* or *Hydrodynamics*, which studies fluids in motion.

Nothing in this world is ideal! This is not a pessimistic thought but simply an interpretation of the term "*ideal*", which originates from the Greek word "ιδέα". It means something that exists only in our mind, an entity that is nonexistent in reality. Given that *real* fluids have physical characteristics (such as, e.g., viscosity) that make the theoretical study of these substances difficult, we invent an idealization of fluids with the generic name *ideal fluid* and with the following properties:

1. An ideal fluid is *absolutely incompressible*. This means that the density of an ideal fluid is the same at all points in the fluid. (As will be seen in Sect. 8.3, this assumption facilitates the derivation of the fundamental equation of Hydrostatics.)
2. An ideal fluid is *absolutely non-viscous* (there is no internal friction within the fluid). This is particularly important in Hydrodynamics since it allows us to use conservation of mechanical energy for the study of fluid motion (this will be discussed in Sect. 8.12).

Many real fluids (e.g., water) have properties that are close to those of an ideal fluid. Note, however, that the characteristics of real fluids are not always undesirable. For

C. J. Papachristou, *Introduction to Mechanics of Particles and Systems*, https://doi.org/10.1007/978-3-030-54271-9_8

example, if fluids were *perfectly* incompressible they would not allow the propagation of elastic waves (such as sound) in their interior. And, of course, you wouldn't enjoy honey so much if it didn't have that familiar thick texture!

8.2   Hydrostatic Pressure

To begin our study of Hydrostatics we consider a liquid at rest in a vessel and we let *ds* be an *elementary* (i.e., infinitesimal) surface located at some point Σ in the fluid (see Fig. 8.1). The element *ds* may be part of the surface of an immersed object or may belong to a fictitious surface within the fluid (that is, a surface consisting of points belonging to the fluid itself). Such an elementary surface may be treated, at least approximately, as a plane surface. We call $d\vec{F}$ the elementary force exerted by the fluid on *ds*.

It is found experimentally that $d\vec{F}$ has the following properties:

1. It is independent of the nature of the surface *ds*. That is, the force exerted by the fluid on the element *ds* does not depend on the molecular composition of *ds* (i.e., on the material of which *ds* is made).
2. The direction of $d\vec{F}$ is always *normal* to *ds*, regardless of the orientation of *ds*. This is a consequence of the absence of internal frictional forces within the ideal fluid and of the fact that the fluid is at rest.
3. The magnitude dF of $d\vec{F}$ is independent of the orientation of *ds* (that is, dF does not change if we rotate *ds* in any direction while leaving the location of *ds* fixed at point Σ). As is found, dF depends only on the location of Σ in the fluid and, for infinitesimal *ds*, it is proportional to the area of *ds* (this infinitesimal area will also be denoted *ds*).

This last remark leads us to the definition of *hydrostatic pressure* P at a point Σ in the fluid:

$$P = \frac{dF}{ds} \quad \Leftrightarrow \quad dF = P\,ds \qquad (8.1)$$

We note the following:

1. In general, P is a function of the location of point Σ in the fluid.
2. P does not depend on the orientation of *ds*; therefore, P is independent of the orientation of $d\vec{F}$. We conclude that P is a *scalar* quantity.

Fig. 8.1 Normal force on an elementary surface *ds* inside a liquid at rest

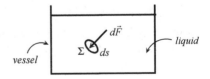

Fig. 8.2 Horizontal surface
S inside a fluid at rest

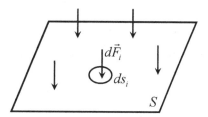

3. P is defined for a particular point in the fluid; it *cannot* be defined as a sum over a set of points. Thus it is meaningless to speak of a "total pressure" on a surface, as we never speak, for example, of a "total temperature" or a "total density" of the air in a room. On the contrary, we *may* define the total force on a finite surface S in a fluid as the vector sum of all elementary forces $d\vec{F}$ exerted by the fluid on the various elementary surfaces ds that make up S.

As verified by experiment,

the hydrostatic pressure is constant over a horizontal surface in a fluid at rest; that is, all points of this surface have the same pressure.

This means that, in a fluid at rest, hydrostatic pressure varies only in the *vertical* direction (i.e., in the direction of the gravitational field of the Earth). In particular, the constant pressure over the *free surface* of the fluid (which surface is always horizontal) is equal to the atmospheric pressure P_0.

Let us now consider a *horizontal surface* of total area S inside a fluid at rest (Fig. 8.2). We partition S into a huge number of elementary surfaces ds_i:

$$S = \sum_i ds_i \ .$$

The total force $\vec{F}$ exerted by the fluid on S is *normal* to S and its magnitude F is the sum of magnitudes dF_i of all elementary forces $d\vec{F}_i$ exerted normally on the corresponding surface elements ds_i:

$$F = \sum_i dF_i \ .$$

But, $dF_i = P ds_i$, where P is the *constant* pressure on S, same for all elementary surfaces ds_i. Hence,

$$F = \sum_i P ds_i = P \sum_i ds_i \quad \Rightarrow$$

$$F = PS \quad \Leftrightarrow \quad P = \frac{F}{S} \tag{8.2}$$

Note carefully that (8.2) is only valid for a *horizontal* surface, since it was derived on the assumption that the pressure P has the same value everywhere on S. On the contrary, the *infinitesimal* relation (8.1) for an elementary surface ds is *always* valid, regardless of the orientation of ds.

8.3 Fundamental Equation of Hydrostatics

It was mentioned earlier that the hydrostatic pressure P in a fluid at rest varies only in the vertical direction. We now seek the equation that describes this variation quantitatively.

Consider an ideal liquid of density ρ. Since the fluid is incompressible, the value of ρ is constant over the entire fluid. If dm is the mass and dV is the volume of an elementary quantity of the fluid, then

$$\rho = \frac{dm}{dV} \quad \Leftrightarrow \quad dm = \rho\, dV \tag{8.3}$$

Consider now a fluid element in the shape of a thin horizontal disk of base area A and infinitesimal thickness dy, thus of volume $dV = A dy$ (Fig. 8.3). The weight of the disk is $dw = (dm)g = (\rho dV)g$, or

$$dw = \rho g A\, dy \tag{8.4}$$

We call $P(y)$ and $P(y + dy)$ the (constant) pressures at the horizontal levels at heights y and $y + dy$, respectively, above an arbitrary reference level $y = 0$. The vertical forces on the disk are its weight, dw, and the normal forces from the liquid on the two horizontal surfaces of the disk, $F(y)$ and $F(y + dy)$. The disk is in equilibrium since it is part of a fluid at rest. Thus, the total vertical force on the disk is zero:

$$\sum F_y = 0 \quad \Rightarrow \quad F(y) - F(y + dy) - dw = 0 \ .$$

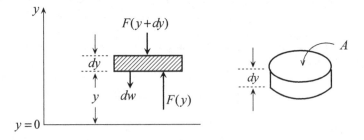

Fig. 8.3 A fluid element in the shape of a thin horizontal disk

But,

$$F(y) = P(y)A , \qquad F(y + dy) = P(y + dy)A .$$

Substituting for dw from (8.4) and eliminating A, we find:

$$P(y + dy) - P(y) = -\rho g \, dy$$

or

$$dP = -\rho g \, dy \qquad (8.5)$$

where dP is the infinitesimal change of pressure corresponding to the change of height dy.

To find the change of pressure $\Delta P = P_2 - P_1$ as we move from a height y_1 to another height y_2 (i.e., as we move a vertical distance $\Delta y = y_2 - y_1$) we integrate (8.5) from y_1 to y_2, taking into account that the density ρ is constant:

$$\int_{P_1}^{P_2} dP = - \int_{y_1}^{y_2} \rho g \, dy = -\rho g \int_{y_1}^{y_2} dy \quad \Rightarrow$$

$$\boxed{P_2 - P_1 = -\rho g \, (y_2 - y_1) \quad \Leftrightarrow \quad \Delta P = -\rho g \, \Delta y} \qquad (8.6)$$

Note that the pressure *decreases* ($\Delta P < 0$) as the height increases ($\Delta y > 0$). Equation (8.6) is often called the *fundamental equation of Hydrostatics*.

Instead of the height y that increases in the *upward* direction, we often refer to the *depth* h below the free surface of the fluid, which increases in the *downward* direction. Then, $dh = -dy$, and the infinitesimal relation (8.5) is rewritten as

$$dP = \rho g \, dh \qquad (8.7)$$

while the fundamental equation (8.6) takes on the form

$$\boxed{P_2 - P_1 = \rho g \, (h_2 - h_1) \quad \Leftrightarrow \quad \Delta P = \rho g \, \Delta h} \qquad (8.8)$$

Note that the pressure now *increases* with depth ($\Delta P > 0$ when $\Delta h > 0$).

As an application, let us determine the pressure P at a point Σ at depth h below the free surface of the liquid (Fig. 8.4). Of course, *all* points at the same depth will have the same pressure.

The pressure at the free surface, where $h = 0$, is the atmospheric pressure P_0. Substituting $h_1 = 0$, $P_1 = P_0$ and $h_2 = h$, $P_2 = P$ in (8.8), we have:

$$P - P_0 = \rho g \, (h - 0) \quad \Rightarrow$$

Fig. 8.4 A point Σ at depth h below the free surface of a liquid at rest

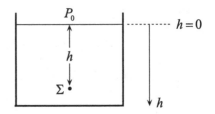

$$\boxed{P = P_0 + \rho g h} \tag{8.9}$$

Relation (8.9) is an alternative, equivalent form of the fundamental equation (8.8). [*Exercise*: By applying (8.9) at two depths h_1 and h_2, recover Eq. (8.8).] Note that the atmospheric pressure P_0 adds to the pressure caused by the liquid alone at point Σ. This is a consequence of *Pascal's principle*, to be examined in Sect. 8.6. More on the atmospheric pressure will be said in Appendix E.

8.4 Units of Pressure

Hydrostatic pressure is expressed in a variety of units, depending on the application of interest. The following units are commonly used in Physics.

1. In the S.I. system of units (*m*, *kg*, *s*) the unit of pressure is the *Pascal* (*Pa*):

$$1\,Pa = 1\,\frac{N}{m^2} \quad \text{where} \quad 1\,N = 1\,kg \cdot m \cdot s^{-2}\,.$$

In the *cgs* system (*cm*, *g*, *s*) the unit of pressure is

$$1\,\frac{dyn}{cm^2} \quad \text{where} \quad 1\,dyn = 1\,g \cdot cm \cdot s^{-2}\,.$$

Given that $1\,kg = 10^3\,g$ and that $1\,m = 10^2\,cm$, we find that $1\,N = 10^5\,dyn$, and therefore,

$$1\,Pa = 10\,\frac{dyn}{cm^2}\,.$$

2. The *Bar* unit is defined as follows:

$$1\,Bar = 10^6\,\frac{dyn}{cm^2} = 10^5\,Pa\,.$$

Hence,

$$1 \frac{dyn}{cm^2} = 10^{-6} \, Bar = 1 \, \mu Bar \; .$$

3. As 1 *Torr* or 1 *mmHg* we define the pressure exerted at the base of a column of mercury (Hg) of height 1 *mm* = 0.1 *cm*. Thus, 1 *Torr* is the pressure change ΔP that relation (8.8) will yield by putting $\rho = 13.6 \, g/cm^3$ (density of Hg), $g = 9.8 \, m/s^2$ and $\Delta h = 0.1 \, cm$:

$$1 \, Torr = (13.6 \, \frac{g}{cm^3}) \times (980 \, \frac{cm}{s^2}) \times (0.1 \, cm)$$

$$= 1332.8 \, \frac{dyn}{cm^2} = 1332.8 \, \mu Bar \; .$$

4. An *atmosphere* (1 *atm*) is defined as the pressure exerted at the base of a column of Hg of height 76 *cm* = 760 *mm*:

$$1 \, atm = 760 \, mmHg = 760 \, Torr = 760 \times 1332.8 \, \mu Bar$$
$$\simeq 1.01 \, Bar \; .$$

This is equal to the standard atmospheric pressure P_0 at sea level:

$$P_0 = 1 \, atm = 760 \, Torr \, (\text{sea level, } 20 \, ^\circ C) \, .$$

Application: Variation of hydrostatic pressure inside the sea

It is known empirically that the hydrostatic pressure inside the sea increases approximately by 1 *atm* for every extra 10 meters of depth. This can be verified numerically by using the fundamental equation (8.8) with $\rho = 1.03 \, g/cm^3$ (density of sea water), $g = 9.8 \, m/s^2$ and $\Delta h = 10 \, m$. We have:

$$\Delta P = \rho g \, \Delta h = (1.03 \, \frac{g}{cm^3}) \times (980 \, \frac{cm}{s^2}) \times (10^3 \, cm) = 1.03 \times 9.8 \times 10^5 \, \frac{dyn}{cm^2}$$

$$= 1.03 \times 9.8 \times 10^{-1} \, Bar = 1.03 \times 9.8 \times 10^{-1} \times \frac{1}{1.01} \, atm \quad \Rightarrow$$

$$\Delta P = 0.9994 \, atm \simeq 1 \, atm \text{ for every extra 10 meters of depth.}$$

Thus, by taking into account that the atmospheric pressure is $P_0 = 1 \, atm$, the pressure at depth h in the sea is found to be

$$P = \left(1 + \frac{h}{10 \, m} \right) atm \; (h \text{ in meters}).$$

Question: What is the pressure at the location where the *Titanic* rests? ($h = 4 \, km$)

8.5 Communicating Vessels

Consider the following experiment: Two vessels of the same height but of different width communicate with each other by a narrow tube whose ends are firmly attached to the vessels at points close to their bottoms. We call this structure a *system of communicating vessels*. We put this system on a table and slowly but steadily pour water into both containers simultaneously. Which vessel will be filled up first?

You may be inclined to answer, "*the narrower one, since the free surface of the water will reach the opening of the vessel sooner than the water in the wider vessel*". If we actually perform the experiment, however, we will realize that the two containers fill up *simultaneously*! This is a consequence of the *principle of communicating vessels*, according to which

if two or more vessels communicate with each other, and if all vessels contain the same liquid and are subject to the same external pressure, then, at the state of equilibrium, the free surfaces of the liquid are at the same horizontal level (i.e., raise to the same height) in all vessels.

The theoretical proof of the principle is as follows (see Fig. 8.5; for convenience, only two vessels are drawn). Consider two points A and A' located at different vessels but belonging to the same horizontal plane. The hydrostatic pressure is thus the same at the two points. Call h and h' the heights of the free surfaces of the liquid above A and A', respectively. By relation (8.9),

$$P_A = P_{A'} \quad \Rightarrow \quad P_0 + \rho g h = P_0 + \rho g h' \quad \Rightarrow \quad h = h' \,.$$

Notice that the validity of this result is independent of the geometrical characteristics of the vessels.

There are two cases where the principle of communicating vessels does *not* apply; namely, (a) when the vessels contain two or more liquids that do not mix, and (b) when the external pressures on the free surfaces of the contained liquid are different in different vessels. Let us see two examples:

a. *Vessels containing immiscible liquids of different densities*

The two vessels in Fig. 8.6 contain liquids of densities ρ_1 and ρ_2, where we assume that $\rho_1 > \rho_2$. We call h_1 and h_2 the heights of the free surfaces of the two liquids relative to the horizontal level of their interface (the existence of such an interface

Fig. 8.5 A system of two communicating vessels containing a liquid of density ρ

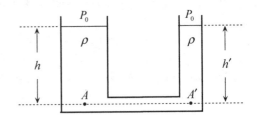

Fig. 8.6 A system of
communicating vessels
containing two liquids that
do not mix

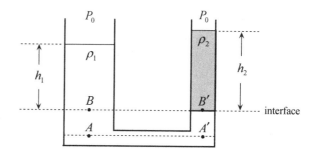

interface

is related to the fact that the liquids do not mix). We consider the points A and A' belonging to the same horizontal plane within the liquid ρ_1. We also consider the point B in the liquid ρ_1, at the same horizontal level with the point B' at the interface of the two liquids. Since A and A' are points in the same liquid, the hydrostatic pressures at these points are equal. By the fundamental equation (8.8),

$$P_A - P_B = \rho_1 g \, (AB) \quad \Rightarrow \quad P_A = P_B + \rho_1 g \, (AB) \;,$$
$$P_{A'} - P_{B'} = \rho_1 g \, (A'B') \quad \Rightarrow \quad P_{A'} = P_{B'} + \rho_1 g \, (A'B') \;.$$

Given that $P_A = P_A{}'$ and $AB = A'B'$, we conclude that $P_B = P_B{}'$. Relation (8.9), then, yields:

$$P_0 + \rho_1 g \, h_1 = P_0 + \rho_2 g \, h_2 \quad \Rightarrow \quad \rho_1 h_1 = \rho_2 h_2 \quad \Rightarrow$$

$$\frac{h_1}{h_2} = \frac{\rho_2}{\rho_1} \;.$$

Thus, if $\rho_1 > \rho_2$, then $h_1 < h_2$.

b. *Vessels at different external pressures*

One end of a U-shaped tube containing mercury (Hg) is connected to a tank containing gas at pressure P, while the other end of the tube is open to the atmosphere, thus subject to the atmospheric pressure P_0 (see Fig. 8.7). We call h the height of the free surface B of Hg at the open end, above the level of the interface between the Hg and the gas in the tank, and we let A be a point a vertical distance h below B (thus, a point at the same horizontal level with the interface).

The hydrostatic pressure at A is equal to the pressure P of the gas in the tank (explain this!) while the pressure at the free surface B of Hg at the open end is equal to the atmospheric pressure P_0. By relation (8.8),

$$P_A - P_B = \rho g h \quad \Rightarrow \quad P - P_0 = \rho g h \;.$$

Fig. 8.7 An open-tube
manometer

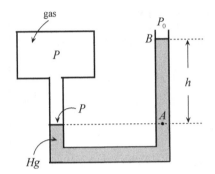

The above-described device is called an *open-tube manometer* and is used for measuring *gauge pressures* $(P - P_0)$. In general, by "gauge pressure" we mean the difference between a variable pressure P and a standard, constant pressure P_0 (such as the atmospheric pressure in our example). The open-tube manometer may be used to measure high pressures exerted by a gas on the walls of a tank.

8.6 Pascal's Principle

Pascal's principle may be stated as follows:

> Every variation of pressure on the free surface of a liquid is felt simultaneously at all points in the liquid. Thus, if the external pressure changes by ΔP, the pressure at all points in the liquid will also change by ΔP.

Proof When the external pressure is P_0, the pressure at a point Σ at depth h below the free surface of the liquid is $P = P_0 + \rho g h$. If the external pressure increases by ΔP, so that its new value is $P_0' = P_0 + \Delta P$, the pressure at Σ will become

$$P' = P_0' + \rho gh = (P_0 + \Delta P) + \rho gh = (P_0 + \rho gh) + \Delta P = P + \Delta P \ .$$

Pascal's principle has a useful practical application in the *hydraulic lever*. In its simplest form, this consists of two communicating cylindrical vessels of cross-sectional areas S_1 and S_2, where $S_1 < S_2$ (see Fig. 8.8). The vessels contain a fluid such as oil or water. At the top of each cylinder there is a piston by which pressure can be exerted on the fluid at both sides. Assume now that we exert a downward force F_1 on the smaller piston. What force F_2 must be applied to the larger piston in order for the system to be in balance?

The smaller piston exerts a pressure $P = F_1/S_1$ on the fluid. According to Pascal's principle, this pressure is transferred to the larger piston, to which is thus exerted an upward force equal to

$$F_2' = P S_2 = \frac{F_1}{S_1} S_2 \ .$$

Fig. 8.8 A simplified form
of a hydraulic lever

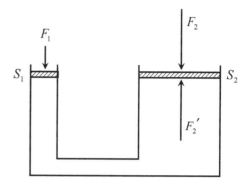

In order for that piston to be in balance we must therefore exert on it a downward force of magnitude $F_2 = F_2'$; that is,

$$F_2 = \frac{S_2}{S_1} F_1 \tag{8.10}$$

We notice that $F_2 > F_1$. Thus, by a small effort (force F_1) we can, e.g., lift a heavy object such as an automobile (force F_2).

8.7 Archimedes' Principle

Archimedes' principle is among the most important principles of Hydrostatics. It is stated as follows:

A body wholly or partially immersed in a liquid is subject to an upward force from the liquid, called <u>buoyant force</u> or <u>buoyancy</u>, which is the resultant of all elementary normal forces exerted by the liquid on the immersed surface of the body. The buoyant force is equal in magnitude to the weight of the fluid displaced by the body, while the line of action of this force passes through the center of gravity of the displaced fluid (<u>center of buoyancy</u>).

The principle is proven theoretically as follows:

Let us call V_d and $\vec{W}_d$ the volume and the weight, respectively, of the fluid displaced by the body. (If the body is wholly immersed in the liquid, V_d equals the volume of the body. If, however, the body is only partially immersed, then V_d is lesser than the body's total volume.) Without loss of generality, we assume that the body is wholly immersed.

Part (a) of Fig. 8.9 shows an instantaneous picture of the immersed body. The word "instantaneous" is related to the fact that, in general, the body is *not* in a state of equilibrium inside the liquid. The buoyant force $\vec{A}$ is the resultant of all elementary forces acting normally on the surface of the body by the liquid.

In part (b) of the figure the body has been removed and has been replaced by liquid of the same volume and shape. The surface of that section of the fluid is now

Fig. 8.9 Instantaneous
picture of an immersed body
(left) and the equivalent
volume of fluid displaced by
the body (right)

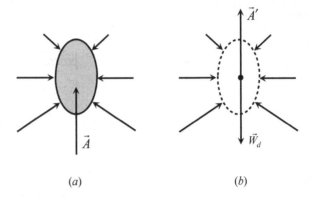

(a)   (b)

subject to a total force $\vec{A}'$ (buoyant force) from the surrounding fluid. The weight $\vec{W}_d$ of this fluid section is equal to the weight of the fluid that had previously been displaced by the body, while the line of action of $\vec{W}_d$ passes through the center of gravity of the displaced fluid.

In contrast to the submerged body, the part of the liquid that replaced the body is in a state of equilibrium since it is a portion of a fluid at rest. Hence,

$$\vec{A}' + \vec{W}_d = 0 \quad \Rightarrow \quad \vec{A}' = -\vec{W}_d \ .$$

Now, the buoyant force on the body is the same as the buoyant force on the part of the fluid replacing the body (i.e., $\vec{A} = \vec{A}'$) since, as mentioned in Sect. 8.2, the elementary forces exerted by a fluid on a surface are independent of the nature of the surface. Thus, finally, the buoyant force exerted by the fluid on the body is

$$\vec{A} = -\vec{W}_d \tag{8.11}$$

The direction of the buoyant force is upward (that is, opposite to the direction of $\vec{W}_d$) while the magnitude of this force is

$$A = W_d = \rho g \, V_d \tag{8.12}$$

where ρ is the density of the liquid.

We note the following:

1. The buoyant force depends only on the volume of the *immersed* part of the body; it is independent of the weight, the density or the chemical composition of the body.
2. For a body that is wholly immersed, the buoyant force on it is independent of the depth at which the body is located inside the liquid.
3. The buoyant force does *not necessarily* pass through the center of gravity of the body, unless the body is homogeneous and is totally immersed, in which case its

center of gravity coincides with the center of buoyancy (center of gravity of the displaced liquid).

Question: A diver claims that he feels the "increasing action of buoyancy" as he dives deeper. What can you tell about his understanding of Hydrostatics? Can you correct his statement in order for it to make some sense?

8.8 Dynamics of the Submerged Body

We submerge a body completely into a liquid (Fig. 8.10) and then let the body free. What will be the subsequent state of motion of the body? As we will now see, this motion depends on the *average* density ρ' of the body in comparison to the density ρ of the liquid.

Let m, V, W be the mass, the volume and the weight, respectively, of the submerged body. The *average density* of the body is defined as

$$\rho' = \frac{m}{V} \quad \Leftrightarrow \quad m = \rho' V \tag{8.13}$$

Hence,

$$W = mg = \rho'g V \tag{8.14}$$

Since the body is fully immersed, $V_d = V$, where V_d is the volume of the fluid displaced by the body. By (8.12), then, the buoyant force on the body has magnitude

$$A = \rho g V \ .$$

In vector form, by taking the positive direction (defined by the direction of the unit vector $\hat{u}$ in Fig. 8.10) upward, we have:

Fig. 8.10 A body of volume V, fully submerged in a liquid of density ρ

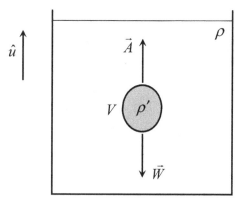

$$\vec{A} = A\,\hat{u} = \rho g\,V\,\hat{u}\ , \qquad \vec{W} = -\,W\,\hat{u} = -\rho'g\,V\,\hat{u}\ .$$

The total force on the body is

$$\vec{F} = \vec{A} + \vec{W} = (A - W)\,\hat{u} = (\rho - \rho')g\,V\,\hat{u} \equiv F\,\hat{u}\ .$$

Note that the direction of $\vec{F}$ depends on the sign of the algebraic value

$$F = A - W = (\rho - \rho')g\,V \tag{8.15}$$

Specifically:

- If $\rho' > \rho$, then $F < 0$ and the body *sinks*.
- If $\rho' = \rho$, then $F = 0$ and the body attains a state of *equilibrium*, fully submerged in the liquid.
- If $\rho' < \rho$, then $F > 0$ and the body *rises* toward the surface of the liquid and finally *floats* in equilibrium, partially submerged in the fluid.

The average density ρ' of a submarine can be varied with the inflow or outflow of seawater, thus becoming larger, smaller or equal to the density ρ of the water. In this way one can achieve diving, surfacing or equilibrium, respectively, of the submarine in the water.

8.9   Equilibrium of a Floating Body

As we have seen, for a body to float partially submerged in a liquid (as shown in Fig. 8.11) the average density ρ' of the body must be *smaller* than the density ρ of the liquid ($\rho' < \rho$). We call V and W the volume and the weight, respectively, of the body, and we let V_d be the volume of the immersed part of the body, equal to the volume of the displaced liquid. In addition to its weight W, the body is subject to the buoyant force A exerted by the fluid on the *immersed* surface of the body. We want to evaluate the fraction of the total volume of the body that is submerged.

Fig. 8.11 A body partially submerged in a liquid of density ρ

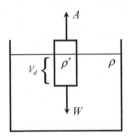

Since the body is in equilibrium, the total force on it is zero. Thus, the buoyant force must exactly balance the weight of the body: $A = W$. But, by (8.12), $A = \rho g V_d$, while by (8.14), $W = \rho' g V$. Hence (by eliminating g),

$$\rho V_d = \rho' V \quad \Rightarrow$$

$$\boxed{\frac{V_d}{V} = \frac{\rho'}{\rho}} \tag{8.16}$$

Thus, e.g., if $\rho' = 3\rho/4$, then $V_d = 3 V/4$. That is, 3/4 of the total volume of the body is immersed, *regardless of the shape or the dimensions* of the body. By applying properties of proportions to (8.16), we find the fraction of the total volume of the body that is *above* the surface of the liquid:

$$\frac{V - V_d}{V} = \frac{\rho - \rho'}{\rho} \tag{8.17}$$

Application: Explain why the captain of the *Titanic* didn't manage to see the iceberg. The density of ice is $\rho' = 0.92$ *gr/cm*^{3}, while that of seawater is $\rho = 1.03$ *gr/cm*^{3}.

Answer: Substituting for ρ and ρ' into (8.17), we find that only 10.68% of the total volume of the iceberg was visible above the surface of the sea.

It should be noted that the condition $A = W$ guarantees the *translational* but not the *rotational* equilibrium of a floating body. What will happen if the body is tipped slightly from its equilibrium position by rotation by a small angle about a horizontal axis passing through the body's center of mass? If the body tends to return to its initial position, the equilibrium is said to be *stable*. If, however, the body tends to depart further from its initial position, the equilibrium is *unstable*. Finally, if the body remains in its new position, the equilibrium is *neutral*.

If the center of gravity C of the body is located *below* the center of buoyancy K (center of gravity of the liquid displaced by the body) the equilibrium is *stable*, since, if the body is rotated slightly with respect to its equilibrium position, its weight $\vec{W}$ and the buoyant force $\vec{A}$ form a *restoring couple* that compels the body to return to its initial position.

It is possible, however, for the equilibrium of a floating body to be stable even if the center of gravity C of the body is located *above* the center of buoyancy K. This depends on the position of C relative to another point M, called the *metacenter*, the location of which is found as follows: When the body is in equilibrium, the axis CK passing through C and K is vertical. We imagine that this axis is firmly attached to the body and thus rotates with it. When the body is deflected from its equilibrium position, a *new* center of buoyancy K' emerges, since the geometry of the displaced fluid changes, in general. The buoyant force $\vec{A}$ now passes through K'. The metacenter M is the point of intersection of the line of action of $\vec{A}$ with the axis CK. One may prove the following:

- If C is located *below M*, the equilibrium is *stable*.
- If C is located *above M*, the equilibrium is *unstable*.
- If C *coincides* with M, the equilibrium is *neutral*.

In ships, the center of gravity C is always higher than the center of buoyancy K but, for relatively small angles of deflection from the vertical, C is located below the metacenter M. Thus the equilibrium of a ship is stable. For angles of deflection larger than a certain limit value, the metacenter M can pass below C, in which case the weight of the ship and the buoyant force form a couple that makes the ship overturn.

In submarines, the metacenter M coincides with the *constant* center of buoyancy K. Thus the equilibrium of a submarine is stable when the center of gravity C is *lower* than K. This can be achieved with the inflow of seawater into suitable tanks.

8.10 Fluid Flow

Having studied the fundamentals of *Hydrostatics* (fluids at rest) we now turn our attention to *Hydrodynamics* (fluids in motion). The motion of a fluid is called *fluid flow*.

It will be helpful to consider that the fluid is composed of a huge number of elementary *fluid particles* (you may visualize them as infinitesimal volume elements) moving in the direction of the flow at each point. The *flow velocity* at a given point at a given time may thus be defined as the velocity of the fluid particle passing through that point at that time.

Since the study of a real flow is often a complicated problem, we will resort again to certain simplifying idealizations.[1] We thus envisage an *ideal flow* having the following characteristics:

1. The fluid is an ideal liquid; thus, it is *incompressible* and *non-viscous*.
2. The flow is *steady*. By this we mean that the flow velocity at any given point is *constant in time* (although it may change from one point to another).
3. The flow is *irrotational*. This means that, at any point in the flow, the passing fluid particle has no angular momentum relative to that point (or, equivalently, relative to this particle's center of mass). The particle thus executes *purely translational* motion.

The path of a fluid particle is called a *streamline*. At any point of a streamline the flow velocity is a vector *tangent* to the line, as shown in Fig. 8.12.

In steady flow, every fluid particle passing through any point A always follows the same streamline (otherwise the flow velocity $\vec{v}_A$ at that point would not be constant in time, since its direction would change within the time interval between the passing of one particle and the passing of another). This means that *streamlines do not cross one another* (they do not intersect).

---

[1] For a discussion of real-fluid flow see, e.g., Chap. 16 of [1].

Fig. 8.12 The flow velocity
is tangential at each point of
a streamline

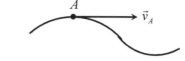

Fig. 8.13 A tube of flow

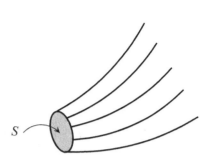

A large aggregate of streamlines forming a bundle of tubular shape is called a *tube of flow*. We can imagine the formation of such a tube as follows: Consider a small plane surface S normal to the streamlines at some location in the flow (see Fig. 8.13). The set of all streamlines passing through the interior as well as through the border of S constitute a tube of flow. In particular, the streamlines passing through the border of S constitute the *boundary* of the tube of flow. The flow velocity at the cross-section S of the tube is defined as the velocity of any fluid particle passing through the center of S (assuming that the tube is narrow enough for the flow velocity to be considered nearly constant over the entire cross-section S).

A tube of flow behaves *like* a real pipe with impenetrable boundaries. Indeed, a fluid particle in the interior or in the exterior of the tube cannot cross the boundary of the tube since, if that happened, the streamline of this particle would cross a streamline belonging to the boundary. Of course, a tube of flow may also possess real, impenetrable boundaries, as happens in the case of a water hose or a water pipe.

8.11 Equation of Continuity

The equation of continuity is the first of two fundamental principles of Hydrodynamics. It is an immediate consequence of the properties of ideal flow and, physically, it expresses conservation of mass (or, equivalently, of volume, in the case of an incompressible fluid).

We consider a tube of flow and two cross-sections of it at points 1 and 2, with corresponding cross-sectional areas A_1 and A_2 (Fig. 8.14). In general, a cross-section of a tube of flow is assumed to be normal to the flow velocity (equivalently, to the central streamline of the tube) at the location of the cross-section. Thus, the flow velocities $\vec{v}_1$ and $\vec{v}_2$ at points 1 and 2 of the tube are perpendicular to the corresponding cross-sections A_1 and A_2.

Fig. 8.14 A tube of flow and
two cross-sections of it

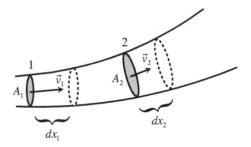

Assume that, within an infinitesimal time interval dt, the fluid particles passing through the cross-section A_1 advance a distance dx_1, while those passing through A_2 advance a distance dx_2. Thus, the fluid volumes passing through the two cross-sections within the time interval dt are

$$dV_1 = A_1 dx_1, \quad dV_2 = A_2 dx_2 \; .$$

But, since the fluid is incompressible and the boundaries of the tube are impenetrable, the volume of the fluid passing through a cross-section will be the same as the volume passing through any other cross-section, within the same time interval. Hence,

$$dV_1 = dV_2 \; .$$

Furthermore, if v_1 and v_2 are the flow speeds at the two cross-sections, then

$$dx_1 = v_1 dt, \quad dx_2 = v_2 dt \; .$$

We thus have:

$$A_1 v_1 dt = A_2 v_2 dt \Rightarrow$$

$$\boxed{A_1 v_1 = A_2 v_2} \tag{8.18}$$

Given that the points 1 and 2 of the tube are chosen arbitrarily, relation (8.18) is equivalently restated as follows:

$$\boxed{A \cdot v \; = \; constant \; along \; the \; tube \; of \; flow} \tag{8.19}$$

Relations (8.18) and (8.19) are alternate versions of the *equation of continuity* for a tube of flow. Note that the proof of this equation was based on the assumption that the boundaries of the tube are impenetrable, which means that it is impossible for any quantity of fluid to either enter or exit the tube. This implies that the quantity of fluid passing through any cross-section of the tube per unit time must be the same for all cross-sections of the tube.

The product $A \cdot v$ carries a particular physical significance, which becomes evident by noting that

$$A\,v = A\,\frac{dx}{dt} = \frac{dV}{dt}$$

where dV is the volume of fluid passing through the cross-section A within time dt. Thus, the product $A \cdot v$ represents the *fluid volume per unit time* passing through the cross-section A of the tube; it is called the *volume flow rate* (or *volume flux*):

$$\Pi = A\,v = \frac{dV}{dt} = volume\ flow\ rate \tag{8.20}$$

According to the equation of continuity (8.19), *the flow rate is constant along the tube*. That is, the same volume of fluid passes per unit time through every cross-section of the tube, in accordance with a remark made earlier.

8.12 Bernoulli's Equation

Bernoulli's equation, to be proven in Appendix F, is an expression of conservation of mechanical energy for a fluid (here, a liquid). Specifically, it is the analog of relation (6.28) for the case of a mechanical system consisting of elementary fluid particles.

Consider a tube of flow of a liquid of density ρ. Consider also two cross-sections of the tube, of areas A_1 and A_2, at the corresponding points 1 and 2 of the tube (Fig. 8.15). The flow velocities, of magnitudes v_1 and v_2, are normal to the corresponding cross-sections. We call P_1, P_2 the hydrostatic pressures at the two cross-sections, and we

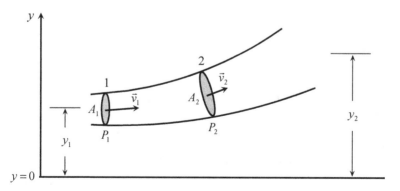

Fig. 8.15 A tube of flow and two cross-sections at different heights above an arbitrary reference level

call y_1, y_2 the heights at which the centers of these cross-sections are located above an arbitrary reference level.

To understand how the pressures P_1 and P_2 are defined, let us consider the section of the tube extending from point 1 to point 2. This section is bounded by the cross-sections A_1 and A_2. Assuming that, at any point of the tube, the hydrostatic pressure is constant (or approximately constant) over the entire cross-section at that point, we write:

$$P_1 = \frac{F_1}{A_1} \ , \qquad P_2 = \frac{F_2}{A_2}$$

where F_1, F_2 are the forces exerted normally on the cross-sections A_1, A_2, respectively, by the fluid surrounding the considered section of the tube. (These forces must be normal to the corresponding cross-sections, given that, in general, every cross-section of the tube moves in a direction perpendicular to itself.)

According to *Bernoulli's equation*, for any two points 1 and 2 of the tube of flow,

$$\boxed{P_1 + \frac{1}{2}\,\rho\,v_1^2 + \rho g y_1 = P_2 + \frac{1}{2}\,\rho\,v_2^2 + \rho g y_2} \qquad (8.21)$$

or, equivalently,

$$\boxed{P + \frac{1}{2}\,\rho\,v^2 + \rho g y = constant \ \ along \ the \ tube \ of \ flow} \qquad (8.22)$$

If we overlook, for a moment, the presence of the pressure P, the left-hand side of (8.22) looks like a total mechanical energy in the gravitational field of the Earth, except that in place of the mass m we now have the density ρ of the fluid. The quantity P here is related to the work required to change this mechanical energy, according to relation (6.28). To be specific, in addition to being subject to the conservative force of gravity, the considered section of the tube is subject to the normal forces F_1, F_2 by the surrounding fluid. The work of these forces is represented by the term P in Bernoulli's equation. This work is responsible for the change of total mechanical energy of the section of the tube (for more details, see Appendix F).

In the trivial case where the flow velocity is zero, the equations of Hydrodynamics must reduce to those of Hydrostatics. For a fluid at rest, we must put $v_1 = v_2 = 0$ into Bernoulli's equation (8.21). This equation then yields:

$$P_2 - P_1 = -\rho g\,(y_2 - y_1) \qquad (8.23)$$

Relation (8.23) is precisely the fundamental equation (8.6) of Hydrostatics.

8.13   Horizontal Flow

We say that a tube of flow is horizontal if the central streamline (which does not have to be rectilinear) lies on a horizontal plane.

All cross-sections of a horizontal tube of flow are centered at the same height y above a horizontal reference level. Thus, by putting $y_1 = y_2$ into Bernoulli's equation (8.21) the y-dependent terms cancel out and the equation reduces to

$$P_1 + \frac{1}{2}\,\rho\,v_1^2 = P_2 + \frac{1}{2}\,\rho\,v_2^2 \tag{8.24}$$

or

$$P + \frac{1}{2}\,\rho\,v^2 = constant \tag{8.25}$$

For a fluid at rest ($v = 0$) it follows from (8.25) that the hydrostatic pressure P is constant along any horizontal path. This conclusion is in perfect agreement with Hydrostatics. Things are different in Hydrodynamics, however, where $v \neq 0$. According to (8.25), in order for the pressure P along a horizontal tube of flow to be constant, the flow speed v must be constant along the tube. On the other hand, by the equation of continuity, the product $A \cdot v$ is constant along the tube, where A is the cross-sectional area of the tube. Hence A must be constant as well. We conclude that

the hydrostatic pressure is constant along a horizontal tube of flow having a constant cross-sectional area.

According to (8.25), in horizontal flow the hydrostatic pressure P increases where the flow speed v decreases. Now, according to the equation of continuity, the flow speed decreases where the cross-sectional area A of the tube increases. Therefore,

in a horizontal tube of flow the hydrostatic pressure increases (decreases) where the cross-sectional area of the tube increases (decreases).

As an application, consider a horizontal pipe of variable cross-sectional area A (Fig. 8.16). Consider two points—say, 1 and 2—of the pipe and let A_1 and A_2 be the cross-sectional areas of the pipe at these points, where $A_1 > A_2$. A liquid of density ρ flows in the pipe. The pressure difference ($P_1 - P_2$) between the two points is

Fig. 8.16 A horizontal pipe of variable cross-sectional area

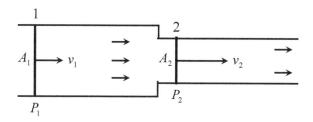

measured and found equal to ΔP. What is the volume flow rate of this horizontal flow?

Let v_1 and v_2 be the flow speeds at the considered two points of the pipe (which pipe constitutes a horizontal tube of flow). We have a system of two equations:

$$A_1 v_1 = A_2 v_2 \quad \text{(equation of continuity)}$$

$$P_1 + \frac{1}{2}\, \rho\, v_1^2 = P_2 + \frac{1}{2}\, \rho\, v_2^2 \quad \text{(Bernoulli's equation)}$$

By the equation of continuity,

$$v_2 = \frac{A_1}{A_2}\, v_1 \tag{8.26}$$

Substituting this result into Bernoulli's equation, we find:

$$P_1 - P_2 \equiv \Delta P = \frac{\rho\,(A_1^2 - A_2^2)}{2 A_2^2}\, v_1^2 \tag{8.27}$$

We notice that $P_1 > P_2$, given that $A_1 > A_2$. By solving (8.27) for v_1 and by using (8.26) for v_2, we have:

$$v_1 = A_2 \left[\frac{2\,\Delta P}{\rho\,(A_1^2 - A_2^2)} \right]^{1/2}, \qquad v_2 = A_1 \left[\frac{2\,\Delta P}{\rho\,(A_1^2 - A_2^2)} \right]^{1/2} \tag{8.28}$$

The volume flow rate is

$$\Pi = A_1 v_1 = A_2 v_2 = A_1 A_2 \left[\frac{2\,\Delta P}{\rho\,(A_1^2 - A_2^2)} \right]^{1/2} \tag{8.29}$$

Reference

1. R. Resnick, D. Halliday, K.S. Krane, *Physics*, vol. 1, 5th ed. (Wiley, 2002)

Appendix A
Composition of Forces Acting in Space

Consider a system of forces $\vec{F}_1 , \vec{F}_2 , \cdots$, acting at various points of space with corresponding position vectors $\vec{r}_1 , \vec{r}_2 , \cdots$, relative to a fixed reference point O that is chosen to be the origin of our coordinate system (x, y, z). These forces are assumed to act on a specific physical system; e.g., on individual members of a system of particles or at certain points of a rigid body. We now ask the question: Under what conditions can the above set of forces be replaced by a single force that will produce the same translational and rotational effects on the physical system?

It is apparent that, if it exists, this force will be the *resultant* $\vec{R}$ of the given system of forces:

$$\vec{R} = \vec{F}_1 + \vec{F}_2 + \cdots = \sum_i \vec{F}_i \tag{A.1}$$

The question now is: *where* exactly must we place $\vec{R}$? If a well-defined answer to this question exists, then the replacement of a set of forces by a single force is possible. The necessary condition for this to be the case is the following:

The torque of the resultant force $\vec{R}$, relative to any point O of space, must be equal to the vector sum of the torques of the component forces $\vec{F}_1 , \vec{F}_2 , \cdots$, relative to O.

Let C be the point of application of $\vec{R}$ (assuming that such a point exists), with position vector $\vec{r}_C$ with respect to O (Fig. A.1). The torques of $\vec{F}_1, \vec{F}_2, \cdots$, with respect to O, are

$$\vec{T}_1 = \vec{r}_1 \times \vec{F}_1 , \quad \vec{T}_2 = \vec{r}_2 \times \vec{F}_2 , \cdots$$

and the total torque of the system of forces, relative to O, is

$$\vec{T} = \vec{T}_1 + \vec{T}_2 + \cdots = \sum_i \vec{T}_i = \sum_i \vec{r}_i \times \vec{F}_i \tag{A.2}$$

© The Editor(s) (if applicable) and The Author(s), under exclusive license
to Springer Nature Switzerland AG 2020
C. J. Papachristou, *Introduction to Mechanics of Particles and Systems*,
https://doi.org/10.1007/978-3-030-54271-9

Fig. A.1 A system of forces
acting at various points of
space; the resultant force is
assumed to apply at C

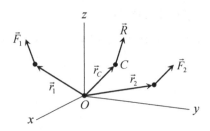

On the other hand, the torque of $\vec{R}$ relative to O is

$$\vec{r}_C \times \vec{R} = \vec{r}_C \times \sum_i \vec{F}_i \, .$$

Thus, the necessary condition stated above is written:

$$\vec{r}_C \times \sum_i \vec{F}_i = \sum_i \vec{T}_i \quad \Leftrightarrow \quad \vec{r}_C \times \vec{R} = \vec{T} \tag{A.3}$$

The question then is whether the vector equation (A.3) has a solution for $\vec{r}_C$, for
given $\vec{R}$ and $\vec{T}$ and for *arbitrary* choice of reference point O.

A particular situation is that in which all forces $\vec{F}_i$ act *at the same point A*, which
has position vector $\vec{r}$. Then $\vec{r}_i = \vec{r}$, for all values of i, and

$$\vec{T} = \sum_i \vec{r}_i \times \vec{F}_i = \sum_i \vec{r} \times \vec{F}_i = \vec{r} \times \sum_i \vec{F}_i = \vec{r} \times \vec{R} \, .$$

Hence, relation (A.3) has the solution $\vec{r}_C = \vec{r}$. We conclude that

a system of forces that are concurrent at a point A may be replaced by their resultant, applied
at the same point A.

Things aren't that simple in the case of non-concurrent forces, since the Eq. (A.3)
may not admit a solution for $\vec{r}_C$ in that case. We can state, however, two *necessary
conditions* for the existence of such a solution:

1. The resultant force $\vec{R}$ must not be zero $\left(\vec{R} \neq 0\right)$. Indeed, if the resultant is zero, Eq.
   (A.3) will either be impossible to solve for $\vec{r}_C$ (if $\vec{T} \neq 0$) or will be indeterminate
   (if $\vec{T} = 0$). For example, in the case of a *couple* $\vec{F}$ and $-\vec{F}$ (see Sect. 7.8) the
   vector $\vec{r}_C$ is not defined, since $\vec{R} = 0$ while $\vec{T} \neq 0$; hence the solution of (A.3) is
   impossible.
2. The total torque $\vec{T}$ of the system, relative to *any* point O, must be normal to the
   resultant force $\vec{R}$. This follows from (A.3) and from the definition of the vector
   product.

We remark that, although the position vector $\vec{r}_C$ of the point of application C of
the resultant force depends on the choice of the reference point O, the location of C

Fig. A.2 A system of
parallel forces and their
resultant

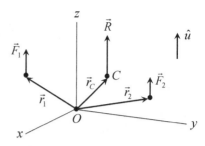

relative to the system of forces must be *uniquely* determined, independently of the
choice of O.

A case where all the above conditions are fulfilled is that of a system of *parallel
forces*:

$$\vec{F}_i = F_i \hat{u}, \quad i = 1, 2, \cdots \tag{A.4}$$

where $\hat{u}$ is a unit vector in the common direction of the forces, and where F_i is the
magnitude of $\vec{F}_i$. The resultant force is

$$\vec{R} = \sum_i \vec{F}_i = \sum_i (F_i \hat{u})$$

or, after factoring out the unit vector $\hat{u}$,

$$\vec{R} = \left(\sum_i F_i \right) \hat{u} \equiv R\hat{u} \tag{A.5}$$

where R is the magnitude of $\vec{R}$. We notice that $\vec{R} \neq 0$; thus the first necessary condition
is satisfied. Furthermore, $\vec{R}$ is parallel to the $\vec{F}_i$ and its magnitude equals the sum of
the magnitudes of the $\vec{F}_i$:

$$R = \sum_i F_i = F_1 + F_2 + \cdots \tag{A.6}$$

Now, let O be an arbitrary point of space, chosen to be the origin of our system
of coordinates (x, y, z). Also, let $\vec{r}_i$ be the position vector, relative to O, of the point
of application of the force $\vec{F}_i$ (Fig. A.2). The total torque of the system of forces,
relative to O, is

$$\vec{T} = \sum_i \vec{T}_i = \sum_i \vec{r}_i \times \vec{F}_i = \sum_i (\vec{r}_i \times F_i \hat{u}) = \sum_i (F_i \vec{r}_i \times \hat{u}) = \left(\sum_i F_i \vec{r}_i \right) \times \hat{u}$$

where we have used a property of the vector product, namely,

$$\vec{A} \times \lambda\vec{B} = (\lambda\vec{A}) \times \vec{B}$$

and we have factored out the unit vector $\hat{u}$. We notice that the total torque $\vec{T}$ is normal to $\hat{u}$, thus also to the total force $\vec{R}$, in accordance with the second necessary condition stated above.

Let C be the point of application of $\vec{R}$, and let $\vec{r}_C$ be the position vector of that point relative to O. In order to determine $\vec{r}_C$ we must solve the vector equation (A.3) for the given $\vec{R}$ and $\vec{T}$:

$$\vec{r}_C \times \vec{R} = \vec{T} \quad \Rightarrow \quad \vec{r}_C \times R\hat{u} = \vec{T} \quad \Rightarrow$$

$$(R\vec{r}_C) \times \hat{u} = \left(\sum_i F_i \vec{r}_i\right) \times \hat{u}.$$

The above equation is satisfied by requiring that

$$R\vec{r}_C = \sum_i F_i \vec{r}_i \quad \Rightarrow$$

$$\vec{r}_C = \frac{1}{R}\sum_i F_i \vec{r}_i = \frac{F_1\vec{r}_1 + F_2\vec{r}_2 + \cdots}{F_1 + F_2 + \cdots} \tag{A.7}$$

The coordinates of the point C, which is called the *center of the parallel forces*, are

$$x_C = \frac{1}{R}\sum_i F_i x_i, \quad y_C = \frac{1}{R}\sum_i F_i y_i, \quad z_C = \frac{1}{R}\sum_i F_i z_i \tag{A.8}$$

We observe that the location of C is independent of the direction of the parallel forces [indeed, notice that (A.7) does not contain $\hat{u}$]. We must now verify that the location of C in space is also independent of the choice of the reference point O. Let us assume, however, that the point of application of $\vec{R}$ does depend on the choice of reference point. Thus, let C and C' be two different points of application, corresponding to the reference points O and O' (Fig. A.3). We call $\vec{r}_C$ and $\vec{r}'_C$ the position vectors of C and C' relative to O and O', respectively, and we call $\vec{r}_i$ and $\vec{r}'_i$ the position vectors of the point of application of $\vec{F}_i$ with respect to O and O' We denote by $\vec{b}$ the vector $\overrightarrow{OO'}$.

Equation (A.7), expressed with respect to both O and O', yields

$$\vec{r}_C = \frac{1}{R}\sum_i F_i \vec{r}_i, \quad \vec{r}'_C = \frac{1}{R}\sum_i F_i \vec{r}'_i$$

Fig. A.3 Hypothetically
different locations C and C'
of the center of a system of
parallel forces, relative to the
reference points O and O'

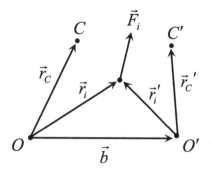

where $\vec{r}_i' = \vec{r}_i - \vec{b}$. Now, by a slight generalization of what was said in Sect. 1.1 regarding the sum of vectors, we have:

$$\overrightarrow{CC'} = \overrightarrow{CO} + \overrightarrow{OO'} + \overrightarrow{O'C'} = -\vec{r}_C + \vec{b} + \vec{r}_C' \Rightarrow$$

$$\overrightarrow{CC'} = -\frac{1}{R}\sum_i F_i \vec{r}_i + \vec{b} + \frac{1}{R}\sum_i F_i \vec{r}_i' = \vec{b} - \frac{1}{R}\sum_i F_i(\vec{r}_i - \vec{r}_i')$$

$$= \vec{b} - \frac{1}{R}\sum_i F_i \vec{b} = \vec{b} - \frac{1}{R}\left(\sum_i F_i\right)\vec{b} = \vec{b} - \frac{1}{R}R\vec{b} = 0$$

by which we conclude that the points C and C' coincide. Hence, the point of application of the total force $\vec{R}$ is independent of the choice of the origin of our coordinate system.

We conclude that

a system of parallel forces is equivalent to a single force, their resultant, located at the center of the parallel forces.

As an application, we now define the *center of gravity* of a system of particles and we demonstrate that this point coincides with the center of mass of the system. Consider a system of particles of masses $m_1, m_2, \ldots$, located at points with position vectors $\vec{r}_1, \vec{r}_2, \cdots$ relative to the origin O of our coordinate system (Fig. A.4). The weights $\vec{w}_1, \vec{w}_2, \cdots$ of the particles constitute a system of parallel forces:

$$\vec{w}_i = m_i g \hat{u} = w_i \hat{u} \tag{A.9}$$

where $w_i = m_i g$ and where $\hat{u}$ is a unit vector perpendicular to the surface of the Earth and directed downward. The total weight of the system of particles is

$$\vec{w} = \sum_i \vec{w}_i = \sum_i (m_i g \hat{u}) = \left(\sum_i m_i\right) g \hat{u} \Rightarrow$$

Fig. A.4 Center of gravity C
of a system of point masses

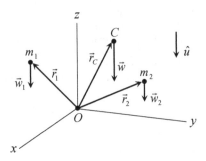

$$\vec{w} = Mg\hat{u} = w\hat{u} \tag{A.10}$$

where M is the total mass of the system, and where $w = Mg$.

The *center of gravity* of a system of particles is the center C of the parallel forces $\vec{w}_i$; that is, the point of application of the total weight $\vec{w}$ of the system. (Note that the location of this point does not have to coincide with that of a particle!) The position of C relative to O is found from Eq. (A.7) by setting $F_i = w_i$ and $R = w$:

$$\vec{r}_C = \frac{1}{w} \sum_i w_i \vec{r}_i = \frac{1}{Mg} \sum_i m_i g \vec{r}_i = \frac{1}{Mg} g \sum_i m_i \vec{r}_i \quad \Rightarrow$$

$$\vec{r}_C = \frac{1}{M} \sum_i m_i \vec{r}_i \tag{A.11}$$

We observe that

the center of gravity of a system of particles coincides with the center of mass of the system

[see Eq. (6.2)]. As shown previously,

the location of this point with respect to the system of particles is uniquely determined; in particular, it does not depend on the choice of the origin of our coordinate system.

The Cartesian coordinates of the center of gravity (equivalently, of the center of mass) of the system are given by the algebraic equations

$$x_C = \frac{1}{M} \sum_i m_i x_i , \quad y_C = \frac{1}{M} \sum_i m_i y_i , \quad z_C = \frac{1}{M} \sum_i m_i z_i \tag{A.12}$$

Appendix B
Some Theorems on the Center of Mass

Consider a system of particles of masses m_i ($i = 1, 2, \ldots$). The total mass of the system is

$$M = m_1 + m_2 + m_3 + \cdots = \sum_i m_i \qquad (B.1)$$

The *center of mass* of the system is defined as the point C with position vector

$$\vec{r}_C = \frac{1}{M}(m_1 \vec{r}_1 + m_2 \vec{r}_2 + \cdots) = \frac{1}{M} \sum_i m_i \vec{r}_i \qquad (B.2)$$

relative to the origin O of our coordinate system (see Fig. 6.1).

Theorem B.1 *The location of C relative to the system of particles does not depend on the choice of the reference point O.*

Proof See Appendix A. For a more direct proof, note first that Eq. (B.2) becomes identical to (A.7) if we put F_i and R in place of m_i and M, respectively; then, simply follow the discussion after Eq. (A.8).

The *total angular momentum* of the system of particles m_i at time t, relative to an arbitrary reference point O, is

$$\vec{L} = \sum_i \vec{L}_i = \sum_i m_i (\vec{r}_i \times \vec{v}_i) \qquad (B.3)$$

© The Editor(s) (if applicable) and The Author(s), under exclusive license
to Springer Nature Switzerland AG 2020
C. J. Papachristou, *Introduction to Mechanics of Particles and Systems*,
https://doi.org/10.1007/978-3-030-54271-9

In particular, the total angular momentum relative to the center of mass C of the system is

$$\vec{L}' = \sum_i m_i (\vec{r}_i' \times \vec{v}_i')$$ (B.4)

where primed quantities are measured with respect to C.

Theorem B.2 *The total angular momentum of the system, with respect to a point O, is the sum of the angular momentum relative to the center of mass ("spin angular momentum") and the angular momentum, relative to O, of a hypothetical particle of mass equal to the total mass of the system, moving with the center of mass ("orbital angular momentum").*

Proof We have:

$$\vec{r}_i = \vec{r}_i' + \vec{r}_C , \quad \vec{v}_i = \vec{v}_i' + \vec{v}_C .$$

Substituting these into (B.3), and using (B.1) and (B.4), we get:

$$\vec{L} = \vec{L}' + M (\vec{r}_C \times \vec{v}_C) + \left[\left(\sum_i m_i \vec{r}_i' \right) \times \vec{v}_C \right] + \left[\vec{r}_C \times \sum_i m_i \vec{v}_i' \right].$$

But, $\Sigma m_i \vec{r}_i' = 0$ and $\Sigma m_i \vec{v}_i' = 0$, since these sums are proportional to the position vector and the velocity, respectively, of the center of mass relative to itself. Thus, finally,

$$\vec{L} = \vec{L}' + M (\vec{r}_C \times \vec{v}_C)$$ (B.5)

Now, suppose O is the origin of an *inertial* reference frame. Let $\vec{F}_i$ be the external force acting on m_i at time t. The *total external torque* acting on the system at this time, relative to O, is given by

$$\vec{T}_{\text{ext}} = \sum_i \vec{r}_i \times \vec{F}_i$$ (B.6)

If we make the assumption that all *internal* forces in the system are *central*, then the following relation exists between the total angular momentum and the total external torque, both quantities measured with respect to O (see Sect. 6.3):

$$\frac{d\vec{L}}{dt} = \vec{T}_{\text{ext}}$$ (B.7)

Equation (B.7) is always valid relative to the origin O of an *inertial* frame. How about its validity relative to the center of mass C of the system? If C moves at constant velocity relative to O, then C is a proper choice of point of reference for the vector relation (B.7). But, what if C is *accelerating* relative to O?

Theorem B.3 *Equation (B.7) is always valid with respect to the center of mass C, even if C is accelerating relative to an inertial frame of reference.*

Proof By differentiating (B.5) with respect to time, by using (B.7) and (B.6), and by taking into account that the total external force on the system is $\vec{F}_{\text{ext}} = M\vec{a}_C$ (see Sect. 6.2), we have:

$$\frac{d\vec{L}}{dt} = \frac{d\vec{L}'}{dt} + M\,(\vec{r}_C \times \vec{a}_C)(+M\,(\vec{v}_C \times \vec{v}_C), \text{ which vanishes}) \quad \Rightarrow$$

$$\vec{T}_{\text{ext}} \equiv \sum_i \vec{r}_i \times \vec{F}_i = \frac{d\vec{L}'}{dt} + (\vec{r}_C \times \vec{F}_{\text{ext}}) \quad \Rightarrow$$

$$\frac{d\vec{L}'}{dt} = \sum_i \vec{r}_i \times \vec{F}_i - \left(\vec{r}_C \times \sum_i \vec{F}_i \right) = \sum_i (\vec{r}_i - \vec{r}_C) \times \vec{F}_i$$

$$= \sum_i \vec{r}_i' \times \vec{F}_i = \vec{T}_{\text{ext}}'$$

where $\vec{T}_{\text{ext}}'$ is the total external torque with respect to the center of mass.

The above theorem justifies using (B.7) to analyze the motion of a rolling body on an inclined plane, even though the axis of rotation passes through the *accelerating* center of mass of the body.

The *total kinetic energy* of the system of particles, relative to an external observer O, is

$$E_k = \sum_i \frac{1}{2} m_i v_i^2 \tag{B.8}$$

The total kinetic energy with respect to the center of mass C is

$$E_k' = \sum_i \frac{1}{2} m_i v_i'^2 \tag{B.9}$$

where, as before, primed quantities are measured relative to C.

Theorem B.4 *The total kinetic energy of the system, relative to an observer O, is the sum of the kinetic energy relative to the center of mass and the kinetic energy, relative to O, of a hypothetical particle of mass equal to the total mass of the system, moving with the center of mass.*

Proof We have:

$$\vec{v}_i = \vec{v}'_i + \vec{v}_C \quad \Rightarrow \quad v_i^2 = \vec{v}_i \cdot \vec{v}_i = v_i'^2 + v_C^2 + 2\vec{v}'_i \cdot \vec{v}_C \, .$$

Substituting this into (B.8), and using (B.1) and (B.9), we get:

$$E_k = E'_k + \frac{1}{2}M v_C^2 + \left(\sum_i m_i \vec{v}'_i \right) \cdot \vec{v}_C \, .$$

But, as noted previously, the sum in the last term vanishes, being proportional to the velocity of the center of mass relative to itself. Thus, finally,

$$E_k = E'_k + \frac{1}{2}M v_C^2 \tag{B.10}$$

Theorem B.5

(a) Consider a system of N particles of masses $m_1, m_2, \ldots, m_N$. Let C be the center of mass of the system. If a new particle, of mass m, is placed at C, the center of mass of the enlarged system of $(N + 1)$ particles will still be at C.

(b) Consider a system of N particles of masses $m_1, m_2, \ldots, m_N$. It is assumed that the location of one of the particles, say of m_N, coincides with the center of mass C of the system. If we now remove this particle from the system, the center of mass of the remaining system of $(N - 1)$ particles will still be at C.

Proof

(a) The total mass of the original system of N particles is $M = m_1 + m_2 + \ldots + m_N$. The center of mass of this system is located at the point C with position vector

$$\vec{r}_C = \frac{1}{M}(m_1 \vec{r}_1 + m_2 \vec{r}_2 + \cdots + m_N \vec{r}_N)$$

relative to some fixed reference point O. For the additional particle, which we name m_{N+1}, we are given that $m_{N+1} = m$ and $\vec{r}_{N+1} = \vec{r}_C$. The total mass of the enlarged system of $(N + 1)$ particles $m_1, m_2, \ldots, m_N, m_{N+1}$ is $M' = M + m$, and the center of mass of this system, relative to O, is located at

$$\vec{r}'_C = \frac{1}{M'}(m_1 \vec{r}_1 + \cdots + m_N \vec{r}_N + m \vec{r}_C) \, .$$

Now, $m_1 \vec{r}_1 + \cdots + m_N \vec{r}_N = M \vec{r}_C$, so that

$$\vec{r}'_C = \frac{1}{M + m}(M \vec{r}_C + m \vec{r}_C) = \vec{r}_C \, .$$

(b) Although this statement is a corollary of part (a) of the theorem, we will treat this as an independent problem. Here we are given that $\vec{r}_N = \vec{r}_C$. Thus,

$$\frac{1}{M}(m_1\vec{r}_1 + \cdots + m_N\vec{r}_N) = \vec{r}_N .$$

The mass of the reduced system of ($N-1$) particles $m_1, m_2, \ldots, m_{N-1}$ is $M' = M - m_N$, while the center of mass of this system is located at

$$\vec{r}'_C = \frac{1}{M'}(m_1\vec{r}_1 + \cdots + m_{N-1}\vec{r}_{N-1}) .$$

But, $m_1\vec{r}_1 + \cdots + m_{N-1}\vec{r}_{N-1} + m_N\vec{r}_N = M\vec{r}_N \Rightarrow$

$$m_1\vec{r}_1 + \cdots + m_{N-1}\vec{r}_{N-1} = (M - m_N)\vec{r}_N = M'\vec{r}_N .$$

Thus, finally,

$$\vec{r}'_C = \frac{1}{M'}M'\vec{r}_N = \vec{r}_N = \vec{r}_C.$$

Appendix C
Table of Moments of Inertia[1]

Solid cylinder of radius R, relative to its axis: $I = \frac{1}{2}MR^2$
Cylindrical shell of radius R, relative to its axis: $I = MR^2$
Circular disk of radius R, relative to a normal axis through its center: $I = \frac{1}{2}MR^2$
Ring of radius R, relative to a normal axis through its center: $I = MR^2$
Solid sphere of radius R, relative to an axis through its center: $I = \frac{2}{5}MR^2$
Spherical shell of radius R, relative to an axis through its center: $I = \frac{2}{3}MR^2$
Thin rod of length l, relative to a normal axis through its center: $I = \frac{1}{12}Ml^2$
Thin rod of length l, relative to a normal axis through one end (*not* a principal axis):
$I = \frac{1}{3}Ml^2$

---

[1] By M we denote the mass of the rigid body.

Appendix D
Principal Axes of Rotation

A *principal axis of rotation* of a rigid body is an axis having the following properties:

- It passes through the center of mass of the body;
- the angular momentum $\vec{L}$ of the body has the same value for all points of reference belonging to that axis, and the direction of $\vec{L}$ is parallel to the axis.

We note that

- every axis of symmetry passing through the center of mass C of the body is a principal axis.

The angular momentum of the rigid body and the total external torque on the body are related by the equation

$$\sum \vec{T} = \frac{d\vec{L}}{dt} \tag{D.1}$$

which is valid relative to a fixed point O of an inertial reference frame, or, relative to the center of mass C of the body (even if C accelerates with respect to O). If the body rotates about a principal axis, the derivative on the right-hand side of (D.1) is a vector that does not depend on the location of the point of reference on that axis. The same must be true, therefore, with regard to the total torque $\sum \vec{T}$ on the left-hand side of (D.1).

As an example, let us consider a thin horizontal disk rotating about a fixed vertical axis passing through the center C of the disk. The axis of rotation is a principal axis since it is an axis of symmetry passing through the center of mass of the disk. Due to that symmetry, for every elementary mass m_i in the disk (see Fig. D.1), moving with velocity $\vec{v}_i$, there is an equal, diametrically opposite mass moving with velocity $-\vec{v}_i$, as seen in Fig. D.2.

The two masses m_i contribute to the total angular momentum of the disk relative to O a quantity equal to the vector sum

C. J. Papachristou, *Introduction to Mechanics of Particles and Systems*, https://doi.org/10.1007/978-3-030-54271-9

Fig. D.1 Two diametrically opposite and equal elementary masses m_i in a horizontal disk rotating about a vertical axis passing through its center

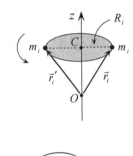

Fig. D.2 The configuration of Fig. D.1 as seen from above (the vertical z-axis is normal to the page and directed toward the reader)

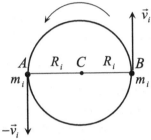

Fig. D.3 A horizontal force applied along the circumference of the disk of Fig. D.1 (here R is the radius of the disk)

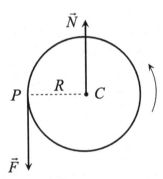

$$\vec{L}_i + \vec{L}_i' = m_i(\vec{r}_i \times \vec{v}_i) + m_i[\vec{r}_i' \times (-\vec{v}_i)] = m_i(\vec{r}_i - \vec{r}_i') \times \vec{v}_i = m_i(\overrightarrow{AB} \times \vec{v}_i).$$

It is easily seen that this vector quantity is in the direction of the z-axis of rotation and, moreover, it does not depend on the location of the point of reference O on this axis. Now, the total angular momentum of the disk, relative to O, is the vector sum of the angular momenta of all *pairs* of diametrically opposite elementary masses. Since the angular momentum of each pair is in the direction of the axis of rotation and does not depend on the location of the reference point O on that axis, the same will be true for the total angular momentum $\vec{L}$ of the disk.

Assume now that we exert a horizontal force $\vec{F}$ on the disk (e.g., by winding a thread around the disk and by pulling the edge of the thread), as shown in Fig. D.3. Since the fixed axis of rotation passes through the center of mass C of the disk, the point C is at rest relative to our inertial reference frame. Accordingly, the total external force on the disk must be zero. In addition to the force $\vec{F}$ that we exert on the disk, there is also the reaction $\vec{N}$ from the pivot at C. Hence,

$$\vec{F} + \vec{N} = 0 \quad \Leftrightarrow \quad \vec{N} = -\vec{F}.$$

We notice that $\vec{F}$ and $\vec{N}$ form a *couple*. As shown in Sect. 7.8, the torque of a couple with respect to O, equal here to the total torque $\Sigma \vec{T}$ on the disk, is

$$\sum \vec{T} = \overrightarrow{CP} \times \vec{F}.$$

We observe that $\Sigma \vec{T}$ does not depend on the location of the reference point O on the axis of rotation, in accordance with a remark made earlier in connection with Eq. (D.1).

In the case of a rolling body, as well as in gyroscopic motion (Sect. 7.12), the principal axis of rotation is not fixed in our inertial reference frame. Indeed, this axis often accelerates with the center of mass C through which it passes. Nevertheless, relation (D.1) is always valid with respect to C, thus with respect to any point of the principal axis. All our previous conclusions, therefore, remain valid.

Appendix E
Variation of Pressure in the Atmosphere

The infinitesimal differential relation

$$dP = -\rho g \, dy \tag{E.1}$$

derived in Sect. 8.3, is valid for *all* fluids, both liquids and gases (remember that, to prove this equation, we did not make any specific assumption regarding the compressibility or not of the fluid). In particular, because of the compressibility of the atmospheric air, the density ρ of the air is not constant but varies with the altitude y above the surface of the Earth.

The air density ρ at a given altitude is approximately proportional to the pressure P of the air at that altitude. Indeed, let us consider a fixed quantity of air, of mass m. Assume that this mass can move to various altitudes, thus be subject to various conditions of pressure and temperature. Consider two specific conditions, called 1 and 2. The volumes occupied by the considered quantity of air are V_1 and V_2, respectively. Applying the equation of state for an (almost) ideal gas to the mass m of air, we have:

$$\frac{P_1 V_1}{T_1} = \frac{P_2 V_2}{T_2} \, .$$

If we make the isothermal approximation $T_1 \cong T_2$, then

$$P_1 V_1 = P_2 V_2 \quad \Rightarrow \quad P_1 \frac{m}{\rho_1} = P_2 \frac{m}{\rho_2} \Rightarrow \frac{P_1}{\rho_1} = \frac{P_2}{\rho_2} \Rightarrow \frac{P}{\rho} = \text{constant} \, .$$

Thus, if $P = P_0$ and $\rho = \rho_0$ are values at sea level ($y = 0$) and if P, ρ are the corresponding values at an arbitrary altitude y, then

$$\frac{P}{\rho} = \frac{P_0}{\rho_0} \quad \Rightarrow \quad \rho = \frac{\rho_0}{P_0} P \, .$$

© The Editor(s) (if applicable) and The Author(s), under exclusive license
to Springer Nature Switzerland AG 2020
C. J. Papachristou, *Introduction to Mechanics of Particles and Systems*,
https://doi.org/10.1007/978-3-030-54271-9

Now, by (E.1),

$$dP = -\left(\frac{\rho_0}{P_0}P\right)gdy \quad \Rightarrow \quad \frac{dP}{P} = -\alpha dy, \quad \text{where we have put } \alpha = \frac{\rho_0 g}{P_0}.$$

By integrating,

$$\int_{P_0}^{P} \frac{dP}{P} = -\alpha \int_{0}^{y} dy \Rightarrow \ln\left(\frac{P}{P_0}\right) = -\alpha y \quad \Rightarrow \boxed{P = P_0 e^{-\alpha y}}.$$

(Show that, then, $\rho = \rho_0 \, e^{-\alpha y}$.) With $\rho_0 = 1.20 \, Kg/m^3 \, (20\,^{\circ}C)$, $P_0 = 1 \, atm$ and $g = 9.8 \, m/s^2$, the value of the constant α is found to be $\alpha = 0.116 \, Km^{-1}$.

Appendix F
Proof of Bernoulli's Equation

Consider a small section of a tube of flow, extending from a to b (Fig. F.1). Let A_1 and A_2 be the cross-sections of the tube at a and b, respectively (we will use the same symbols for the cross-sectional areas). The flow velocities are normal to the corresponding cross-sections. We call v_1, v_2 and P_1, P_2 the flow speeds and the hydrostatic pressures, respectively, at the cross-sections A_1 and A_2. Also, we call y_1 and y_2 the heights at which the centers of the cross-sections are located, with respect to an arbitrary horizontal reference level.

As the considered section of the fluid moves along the tube, within an infinitesimal time interval dt the cross-section A_1 advances from a to a' by a distance $dx_1 = v_1 dt$, while the cross-section A_2 advances from b to b' by $dx_2 = v_2 dt$. Since the ideal fluid

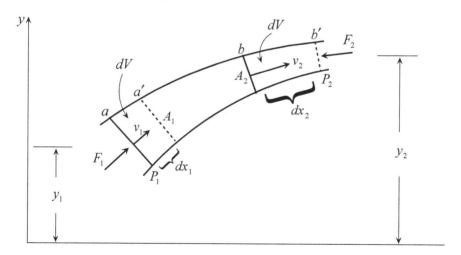

Fig. F.1 A section of a tube of flow, extending from a to b

© The Editor(s) (if applicable) and The Author(s), under exclusive license to Springer Nature Switzerland AG 2020
C. J. Papachristou, *Introduction to Mechanics of Particles and Systems*,
https://doi.org/10.1007/978-3-030-54271-9

is incompressible, the volume of the fluid-section is unchanged in the meantime. Hence, the volumes of fluid contained between a and a', and between b and b', are equal:

$$A_1 dx_1 = A_2 dx_2 = dV \tag{F.1}$$

The same is therefore true with regard to the corresponding masses of fluid:

$$dm_1 = dm_2 = dm = \rho \, dV \tag{F.2}$$

where ρ is the (constant) density of the fluid.

Let F_1, F_2 be the forces exerted normally on the cross-sections A_1, A_2 by the fluid surrounding the considered section of the tube (these forces have to be normal to the corresponding cross-sections, given that, in general, every cross-section of the tube moves in a direction perpendicular to itself). Assuming that, at any point of the tube, the hydrostatic pressure is constant (or approximately constant) over the entire cross-section at that point, we write:

$$F_1 = P_1 A_1, \quad F_2 = P_2 A_2 \tag{F.3}$$

We notice that F_1 is in the direction of the flow, while F_2 is opposite to the flow. This means that F_1 *produces* work (its work is positive) while F_2 *consumes* work (its work is negative). The total work of F_1 and F_2 for an elementary displacement of the section ab of the tube within time dt, is

$$W = F_1 dx_1 - F_2 dx_2 = P_1 A_1 dx_1 - P_2 A_2 dx_2 \quad \Rightarrow$$

$$W = (P_1 - P_2) dV \tag{F.4}$$

where we have used (F.3) and (F.1).

From the discussion in Sect. 6.5, extended here to the case of a continuous medium, it follows that the work W of the forces F_1 and F_2 on the fluid-section ab, within a time dt, equals the change of the total mechanical energy of this section:

$$W = \Delta(E_k + E_p) = \Delta E_k + \Delta E_p \tag{F.5}$$

We have:

$$\Delta E_k = (E_k)_{a'b'} - (E_k)_{ab} = (E_k)_{bb'} - (E_k)_{aa'} \tag{F.6}$$

[Since the flow is steady, the quantity $(E_k)_{a'b}$, for fixed a' and b, is constant in time; thus it is eliminated by subtracting.] Now, by using (F.2), Eq. (F.6) is written:

$$\Delta E_k = \frac{1}{2}(dm)v_2^2 - \frac{1}{2}(dm)v_1^2 = \frac{1}{2}(v_2^2 - v_1^2)dm \quad \Rightarrow$$

$$\Delta E_k = \frac{1}{2}\rho(v_2^2 - v_1^2)dV \tag{F.7}$$

By similar reasoning,

$$\Delta E_p = (E_p)_{bb'} - (E_p)_{aa'} = (dm)gy_2 - (dm)gy_1 = (dm)g(y_2 - y_1) \quad \Rightarrow$$

$$\Delta E_p = \rho g(y_2 - y_1)dV \tag{F.8}$$

Substituting (F.4), (F.7) and (F.8) into (F.5), and eliminating dV, we find:

$$P_1 - P_2 = \frac{1}{2}\rho(v_2^2 - v_1^2) + \rho g(y_2 - y_1) \quad \Rightarrow$$

$$\boxed{P_1 + \tfrac{1}{2}\rho v_1^2 + \rho g y_1 = P_2 + \tfrac{1}{2}\rho v_2^2 + \rho g y_2}$$

This is Bernoulli's equation for an ideal flow.

We note that both the equation of continuity (in the form we wrote it) and Bernoulli's equation can approximately be used for a gas (e.g., atmospheric air) in some limited region of space where the gas density does not vary appreciably from one point to another (the gas may thus be treated as an *almost incompressible fluid* in that region).

Solved Problems

Problems for Chaps. 2–3

1. *The position of a particle moving along the x-axis is given as a function of time by the equation $x = 2t^3 - 3t^2 + 6t - 5$, where x is in meters and t is in seconds. Find (a) the direction of motion at all times and (b) the time intervals during which the motion is accelerated or retarded. Does the particle ever stop?*

   Solution: The algebraic values of velocity and acceleration are

$$v = \frac{dx}{dt} = 6(t^2 - t + 1), \quad a = \frac{dv}{dt} = 6(2t - 1).$$

   Hence,

$$va = 36(t^2 - t + 1)(2t - 1).$$

   We notice that $t^2 - t + 1 > 0$, for all values of t. Thus $v > 0$ for all t, which means that the particle always moves in the positive direction of the x-axis, never stopping (even momentarily). Also, it is easy to verify that $va > 0$ when $t > 0.5$, while $va < 0$ when $t < 0.5$. Hence (cf. Sect. 2.4) the motion is retarded for $t < 0.5$ s and accelerated for $t > 0.5$ s.

2. *A particle moves on the x-axis, having started at time $t = 0$ from the point $x = 0$ with initial velocity $v = v_0$. Find the velocity v and the position x of the particle as functions of time, as well as the velocity as a function of position x, if the acceleration of the particle is given as a function of velocity by (a) $a = -kv$; (b) $a = -kv^2$ (where k is a positive constant).*

   Solution:

   (a) Let $a = -kv$:

C. J. Papachristou, *Introduction to Mechanics of Particles and Systems*,
https://doi.org/10.1007/978-3-030-54271-9

$$\frac{dv}{dt} = -kv \implies \frac{dv}{v} = -kdt \implies \int_{v_0}^{v} \frac{dv}{v} = -k\int_{0}^{t} dt \implies \ln\left(\frac{v}{v_0}\right) = -kt \implies$$

$$\frac{v}{v_0} = e^{-kt} \implies v = v_0 e^{-kt} \tag{1}$$

$$\frac{dx}{dt} = v_0 e^{-kt} \implies dx = v_0 e^{-kt} dt \implies \int_{0}^{x} dx = v_0 \int_{0}^{t} e^{-kt} dt \implies x = \frac{v_0}{k}(1 - e^{-kt}) \tag{2}$$

Eliminating e^{-kt} from (1) and (2), we find:

$$v = v_0 - kx \tag{3}$$

Alternatively, $dv = adt$, $dx = vdt$, and, by dividing these,

$$\frac{dv}{dx} = \frac{a}{v} = -k \implies dv = -kdx \implies \int_{v_0}^{v} dv = -k\int_{0}^{x} dx \implies v - v_0 = -kx \implies \tag{3}$$

(b) Let $a = -kv^2$. Working similarly, show that

$$v = \frac{v_0}{1 + kv_0 t}, \quad x = \frac{1}{k}\ln(1 + kv_0 t), \quad v = v_0 e^{-kx}.$$

3. *A body moves on the x-axis with acceleration $a = (4x - 2)$ m/s ^{2}, where x is the instantaneous position of the body. The initial velocity of the body is $v_0 = 10$ m/s, at the initial location $x_0 = 0$. Find the velocity v of the body as a function of position x.*

Solution: Since the acceleration a is given as a function of x, we use Eq. (2.5):

$$v^2 = v_0^2 + 2\int_{x_0}^{x} adx = 10^2 + 2\int_{0}^{x} (4x - 2)dx \implies v^2 = 4(x^2 - x + 25).$$

We notice that $x^2 - x + 25 > 0$, for all values of x. Hence, $v = \pm 2(x^2 - x + 25)^{1/2}$. We drop the negative sign, since it would yield $v_0 = -10$ for $x = 0$, contrary to the given initial condition. Thus, finally,

$$v = 2(x^2 - x + 25)^{1/2} \text{ m/s}.$$

4. *A particle moves on the circumference of a circle. Its position at time t is given by the equation $s = t^3 + 2t^2$, where s is measured in meters along the circumference and where t is in seconds. The magnitude of the acceleration of the particle at time $t = 2$ s is $a = 16\sqrt{2}$ m/s². Find the radius R of the circle.*

Solution: The magnitude of the velocity of the particle is $v = \frac{ds}{dt} = 3t^2 + 4t$.

We know that $a^2 = a_T^2 + a_N^2 = \left(\frac{dv}{dt}\right)^2 + \left(\frac{v^2}{R}\right)^2$.

Hence, $a^2 = (6t + 4)^2 + \frac{(3t^2 + 4t)^4}{R^2}$.

Setting $t = 2$ and $a = 16\sqrt{2}$, and solving for R, we find: $R = 25\ m$.

5. A particle moves on the xy-plane. Its coordinates x and y are given as functions of time t by the equations: $x = t^2$, $y = (t - 1)^2$, where $t \geq 1$. Find:

   (a) The equation of the trajectory in the form $F(x, y) = constant$;
   (b) the components a_T and a_N of the acceleration, for all t;
   (c) the radius of curvature ρ, for all t.

Solution:

(a) Noting that $x \geq 0$, $y \geq 0$, and taking into account that $t \geq 1$, we write: $\sqrt{x} = t$, $\sqrt{y} = t - 1$. Eliminating t, we find: $F(x, y) \equiv \sqrt{x} - \sqrt{y} = 1$.
(b) We have:

$$v_x = \frac{dx}{dt} = 2t, \quad v_y = \frac{dy}{dt} = 2(t - 1), \quad a_x = \frac{dv_x}{dt} = 2, \quad a_y = \frac{dv_y}{dt} = 2.$$

If $|\vec{v}| = v$ and $|\vec{a}| = a$,

$$a^2 = a_x^2 + a_y^2 = 8, \quad v^2 = v_x^2 + v_y^2 = 8t^2 - 8t + 4.$$

Hence, $v = 2(2t^2 - 2t + 1)^{1/2}$. Note that $v > 0$ for all t.
Then, $a_T = \frac{dv}{dt} = \frac{4t - 2}{(2t^2 - 2t + 1)^{1/2}}$.

For a_N we cannot make direct use of the relation $a_N = v^2/\rho$, since we do not yet know ρ. We thus proceed as follows:

$$a_T^2 + a_N^2 = a^2 = 8 \Rightarrow a_N^2 = 8 - a_T^2 = \frac{4}{2t^2 - 2t + 1} \quad \Rightarrow \quad a_N = \frac{2}{(2t^2 - 2t + 1)^{1/2}}.$$

(c) $a_N = \frac{v^2}{\rho} \Rightarrow \rho = \frac{v^2}{a_N} = 2(2t^2 - 2t + 1)^{3/2}$.

6. Study the **motion of a projectile** (see figure). Find:

   (a) The time t_A for the projectile to reach the highest point A of the trajectory;
   (b) the maximum height h;
   (c) the time t_B at which the projectile hits the ground at point B;
   (d) the total horizontal distance $R = OB$; show that, for a given initial speed v_0, R is a maximum when the angle α is equal to $45°$.
   (e) Show that the speed at B is the same as the initial speed at O; also show that the angles α and β are equal.
   (f) Find the equation of the trajectory, in the form $y = f(x)$.

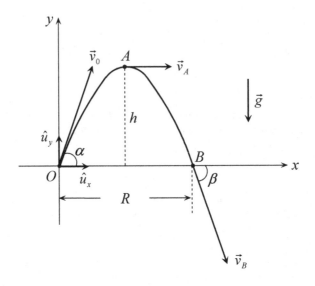

Fig. Problem 6

Solution: We assume that the projectile is ejected from O at time $t = 0$, with initial velocity $\vec{v}_0$. The projectile moves with constant acceleration $\vec{a} = \vec{g} = -g\hat{u}_y$; thus, its motion takes place in a constant vertical plane defined by the vectors of the initial velocity and the acceleration (cf. Sect. 2.5). Consider an arbitrary point of the trajectory, with position vector $\vec{r}$, and let $\vec{v}$ be the velocity of the projectile at that point. The vectors $\vec{r}$ and $\vec{v}$ are given as functions of time by the equations (Sect. 2.5)

$$\vec{v} = \vec{v}_0 + \vec{a}t, \quad \vec{r} = \vec{v}_0 t + \frac{1}{2}\vec{a}t^2$$

or, in component form,

$$v_x\hat{u}_x + v_y\hat{u}_y = (v_{0x}\hat{u}_x + v_{0y}\hat{u}_y) - gt\hat{u}_y,$$
$$x\hat{u}_x + y\hat{u}_y = (v_{0x}\hat{u}_x + v_{0y}\hat{u}_y)t - \frac{1}{2}gt^2\hat{u}_y.$$

By equating the coefficients of $\hat{u}_x$ and $\hat{u}_y$ on the two sides of each equation, we obtain the equations of motion of the projectile in the xy-plane:

$$v_x = v_{0x} = const., \; v_y = v_{0y} - gt$$
$$x = v_{0x}t, \qquad\qquad y = v_{0y}t - \tfrac{1}{2}gt^2$$

where

$$v_{0x} = v_0 \cos\alpha, \quad v_{0y} = v_0 \sin\alpha \quad (v_0 = |\vec{v}_0|).$$

(Note that the motion is uniform in the horizontal direction and uniformly accelerated in the vertical direction.)

(a) At the highest point A, $v_y = 0 \Rightarrow v_{0y} - gt_A = 0 \Rightarrow t_A = \frac{v_{0y}}{g} = \frac{v_0 \sin \alpha}{g}$.

(b) $h = y_A = v_{0y}t_A - \frac{1}{2}gt_A^2 \Rightarrow h = \frac{v_{0y}^2}{2g} = \frac{v_0^2 \sin^2 \alpha}{2g}$.

(c) $y_B = 0 \Rightarrow v_{0y}t_B - \frac{1}{2}gt_B^2 = 0 \Rightarrow t_B = \frac{2v_{0y}}{g} = 2t_A$.

(d) $R = x_B = v_{0x}t_B = \frac{2v_{0x}v_{0y}}{g} = \frac{2v_0^2 \sin \alpha \cos \alpha}{g} \Rightarrow R = \frac{v_0^2 \sin 2\alpha}{g}$.

We notice that $R = \max$ when $\sin 2\alpha = 1$ or $\alpha = 45°$.

(e) At the return point B, $v_x = v_{0x}$, $v_y = v_{0y} - gt_B = v_{0y} - 2v_{0y} = -v_{0y}$.

Thus, $|\vec{v}_B|^2 = v_x^2 + v_y^2 = v_{0x}^2 + v_{0y}^2 = v_0^2 \Rightarrow |\vec{v}_B| = v_0 = |\vec{v}_0|$.

Moreover, $\tan \beta = \frac{|v_y|}{|v_x|} = \frac{v_{0y}}{v_{0x}} = \tan \alpha \Rightarrow \beta = \alpha$.

(f) $x = v_{0x}t \Rightarrow t = \frac{x}{v_{0x}} = \frac{x}{v_0 \cos \alpha}$. Then,

$$y = v_{0y}t - \frac{1}{2}gt^2 \Rightarrow y = -\frac{g}{2v_0^2 \cos^2 \alpha}x^2 + (\tan \alpha)x$$

which is an equation of the form $y = \kappa\, x^2 + \lambda\, x$ (equation of a parabola).

7. *Every body moving in the air "feels" a frictional force $\vec{f}$ (air resistance) the magnitude of which is proportional to the speed of the body, while the direction of this force is opposite to that of the velocity: $\vec{f} = -k\vec{v}$ (where k is a positive constant). Assume that we let a body of mass m fall under the action of gravity with zero initial velocity. (a) Find the velocity of the body as a function of time. Does the body ever stop? (b) Find the maximum value of the velocity during the free fall.*

Solution: The body is subject to two forces (see figure); namely, its weight $\vec{w}$ and the air resistance $\vec{f}$. We write:

$$\vec{w} = mg\hat{u}, \quad \vec{f} = -kv\hat{u}$$

where v is the speed of the body. The total force on the body is

$$\vec{F} = \vec{w} + \vec{f} = (mg - kv)\hat{u}.$$

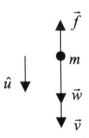

Fig. Problem 7

The acceleration of the body is

$$\vec{a} = \frac{d\vec{v}}{dt} = \frac{d}{dt}(v\hat{u}) = \frac{dv}{dt}\hat{u}.$$

By Newton's law, $\vec{F} = m\vec{a}$, we have (after eliminating $\hat{u}$):

$$mg - kv = m\frac{dv}{dt} \quad \Rightarrow \quad \frac{dv}{dt} = g - \frac{k}{m}v \quad \Rightarrow \quad \frac{dv}{g - \frac{k}{m}v} = dt \quad \Rightarrow$$

$$\int_0^v \frac{dv}{g - \frac{k}{m}v} = \int_0^t dt \quad \Rightarrow \quad -\frac{m}{k}\left[\ln\left(g - \frac{k}{m}v\right)\right]_0^v = t \quad \Rightarrow$$

$$\ln\left[\frac{1}{g}\left(g - \frac{k}{m}v\right)\right] = -\frac{kt}{m} \quad \Rightarrow \quad v = \frac{mg}{k}\left(1 - e^{-\frac{kt}{m}}\right).$$

We notice that $v > 0$ for all $t > 0$. Thus the velocity never vanishes; hence the body never stops. Also, as $t \to \infty$, the speed approaches a maximum value

$$v\text{max} = \frac{mg}{k}.$$

8. *A container of weight $w_\delta = 5$ N has a total capacity of $w = 50$ N of water. The container is placed on a scale and water falls into it from a height $h = 10$ m, at a rate of $\lambda = 0.5$ kg/s. Find the indication of the scale at the moment when the container is exactly half-full. ($g = 9.8$ m/s^2)*

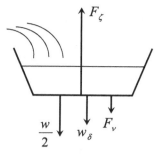

Fig. Problem 8

Solution: The container is in balance under the action of four forces (see figure); namely, its weight w_δ, the weight $w/2$ of the contained water, the force F_ν exerted by the water that falls from height h, and the force F_ζ exerted by the scale. By the action-reaction law, the container exerts on the scale a downward force of magnitude F_ζ; it is precisely this force that corresponds to the indication of the scale. The condition for equilibrium is

$$F_\zeta = w_\delta + \frac{w}{2} + F_\nu \tag{1}$$

To find F_ν we work as follows: Let dm be the mass of water added to the container within an infinitesimal time interval dt. The mass dm falls in the container at time t with a downward velocity $\vec{v}$ and then, within the interval dt, it becomes embodied in the rest of the water and finally balances into it. The momentum of dm at times t and $t + dt$ is

$$\vec{p}(t) = (dm)\vec{v}, \quad \vec{p}(t + dt) = (dm) \cdot 0 = 0.$$

Thus,

$$d\vec{p} = \vec{p}(t + dt) - \vec{p}(t) = -(dm)\vec{v}.$$

The force exerted on dm by the rest of the system is

$$\frac{d\vec{p}}{dt} = -\frac{dm}{dt}\vec{v} = -\lambda\vec{v}$$

where use has been made of the fact that the rate dm/dt at which water falls into the container is λ. By the action-reaction law, dm exerts on the rest of the system a downward force of magnitude $F_\nu = \lambda v$. Now, dm executes free fall from a height h, with no initial velocity. The velocity with which dm falls on the container is

$$v = \sqrt{2gh} \quad \text{(show this!)}.$$

Hence,

$$F_v = \lambda v = \lambda \sqrt{2gh} \tag{2}$$

By (1) and (2), and by making numerical substitutions, we finally have:

$$F_\zeta = w_\delta + \frac{w}{2} + \lambda \sqrt{2gh} = 37\,N \ .$$

9. *A chain of weight w = 2 N is kept vertical from its upper end, while its lower end touches the floor. The chain is then allowed to fall freely. Find the force exerted on the floor at the moment when exactly half the chain has fallen.*

Solution: We basically work as in Problem 8. The total force on the floor at the considered moment is

$$F = \frac{w}{2} + F_\alpha \tag{1}$$

where F_α is the force exerted on the floor due to the falling of the chain. Let *dm* be the mass of a small fraction of the chain, located at the center of the chain. This fraction touches the floor at time *t*, having fallen from height $h = L/2$, where *L* is the total length of the chain. Within time *dt*, the fraction finally comes to rest on the floor. The momentum of *dm* at times *t* and $t + dt$ is

$$\vec{p}(t) = (dm)\vec{v}, \ \vec{p}(t + dt) = (dm) \cdot 0 = 0 \quad \Rightarrow$$
$$d\vec{p} = \vec{p}(t + dt) - \vec{p}(t) = -(dm)\vec{v}$$

where $\vec{v}$ is the velocity with which *dm* falls on the floor, of magnitude

$$v = \sqrt{2gh} = \sqrt{2g\frac{L}{2}} = \sqrt{gL} \tag{2}$$

The upward force exerted on *dm* by the floor is

$$\frac{d\vec{p}}{dt} = -\frac{dm}{dt}\vec{v} \ .$$

By the action-reaction law, then, *dm* exerts a downward force on the floor, of magnitude

$$F_\alpha = \frac{dm}{dt}v \tag{3}$$

Let *dl* be the length of the section *dm*, and let *M* be the total mass of the chain. If $\rho = M/L$ is the linear density of the chain, then

$$dm = \rho dl = \frac{M}{L}dl \Rightarrow \frac{dm}{dt} = \frac{M}{L}\frac{dl}{dt} = \frac{M}{L}v \qquad (4)$$

since the touchdown speed v of dm is equal to dl/dt. By (3) and (4), and by using (2), we have:

$$F_\alpha = \frac{M}{L}v^2 = \frac{M}{L}gL = Mg = w$$

where w is the weight of the chain. Equation (1) finally yields:

$$F = \frac{w}{2} + w = \frac{3w}{2} = 3\,N.$$

10. *A train is moving on a horizontal rectilinear track. The string of a pendulum suspended from the ceiling of the train makes an angle θ with the vertical. The train carries a box placed on the floor of a compartment. Determine the minimum coefficient of static friction between the box and the floor of the train, in order that the box will not slide on the floor.*

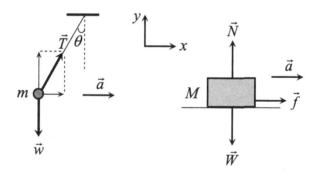

Fig. Problem 10

Solution: Let m be the mass of the pendulum and let M be the mass of the box. The forces on m are its weight $\vec{w}$ and the tension $\vec{T}$ of the string, while those on M are its weight $\vec{W}$, the normal force $\vec{N}$ from the floor of the train, and the static friction $\vec{f}$ (explain why its direction is to the right in the figure). Both bodies move with the acceleration $\vec{a}$ of the train. Newton's law for these bodies is written:

$$\vec{T} + \vec{w} = m\vec{a}, \quad \vec{N} + \vec{W} + \vec{f} = M\vec{a}.$$

Taking x- and y-components, we have:

$$T\sin\theta = ma \qquad (1)$$

$$T\cos\theta - mg = 0 \quad \Rightarrow \quad T\cos\theta = mg \tag{2}$$

$$f = Ma \tag{3}$$

$$N - Mg = 0 \quad \Rightarrow \quad N = Mg \tag{4}$$

Dividing (1) by (2)

$$\Rightarrow \quad \tan\theta = \frac{a}{g} \quad \Rightarrow \quad a = g\tan\theta \tag{5}$$

Dividing (3) by (4) and using (5) $\Rightarrow$

$$\frac{f}{N} = \frac{a}{g} = \tan\theta \quad \Rightarrow \quad f = N\tan\theta \tag{6}$$

Now,

$$f \le f_{max} = \mu N \overset{(6)}{\Rightarrow} N\tan\theta \le \mu N \quad \Rightarrow$$
$$\mu \ge \tan\theta \quad \Leftrightarrow \quad \mu_{min} = \tan\theta .$$

11. *Two masses m_1 and m_2 are connected by a light string and through a pulley, as shown in the figure. (a) Find the minimum coefficient of static friction μ_{min} between m_1 and the horizontal surface, in order for the system to be in equilibrium. (b) For a coefficient of friction $\mu < \mu_{min}$, find the acceleration of the two masses, as well as the tension of the string.*

Solution: We draw the forces separately for each mass. We call $\vec{T}$ the tension of the string; $\vec{f}$ the frictional force between m_1 and the horizontal surface; $\vec{w}_1$ and $\vec{w}_2$ the weights of the two objects; and $\vec{N}$ the normal force on m_1 by the horizontal plane. Note that the magnitude T of the tension of the string is constant along the string and is not affected by the pulley, since the string is not wound around the pulley but simply glides on it without friction.

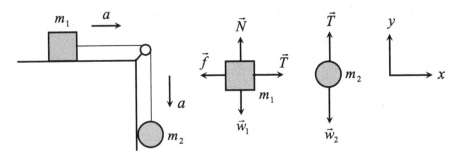

Fig. Problem 11

(a) The system is in balance. Hence, $\sum \vec{F} = 0 \Leftrightarrow \sum F_x = \sum F_y = 0$, for each mass separately. We thus have:

$$T - f = 0 \quad \Rightarrow \quad T = f \tag{1}$$

$$N - w_1 = 0 \quad \Rightarrow \quad N = w_1 = m_1 g \tag{2}$$

$$T - w_2 = 0 \quad \Rightarrow \quad T = w_2 = m_2 g \tag{3}$$

From (1) and (3) $\Rightarrow$

$$f = m_2 g \tag{4}$$

But,

$$f \leq f_{\max} = \mu N \stackrel{(2),(4)}{\Rightarrow} m_2 g \leq \mu m_1 g \quad \Rightarrow$$
$$\mu \geq \frac{m_2}{m_1} \quad \Leftrightarrow \quad \mu_{\min} = \frac{m_2}{m_1}.$$

(b) For $\mu < \mu_{\min}$, the two masses move with accelerations of the same magnitude, a. By applying Newton's law:

$$\sum \vec{F} = m\vec{a} \quad \Leftrightarrow \quad \sum F_x = ma_x, \quad \sum F_y = ma_y,$$

for each mass separately, we have:

$$T - f = m_1 a \quad \Rightarrow \quad T = f + m_1 a \tag{5}$$

$$N - w_1 = 0 \quad \Rightarrow \quad N = w_1 = m_1 g \tag{6}$$

$$T - w_2 = -m_2 a \quad \Rightarrow \quad T = w_2 - m_2 a = m_2 g - m_2 a \tag{7}$$

The friction f is now kinetic, with coefficient μ. Thus,

$$f = \mu N \stackrel{(6)}{\Rightarrow} f = \mu m_1 g.$$

Then, from (5) $\Rightarrow$

$$T = \mu m_1 g + m_1 a \tag{8}$$

From (7) and (8) $\Rightarrow$ $\mu m_1 g + m_1 a = m_2 g - m_2 a$ $\Rightarrow$

$$a = \frac{m_2 - \mu m_1}{m_1 + m_2} g \qquad (9)$$

(Note that $a > 0 \Leftrightarrow m_2 - \mu m_1 > 0 \Leftrightarrow \mu < \frac{m_2}{m_1} = \mu_{min}$.)
Substituting (9) into (7) $\Rightarrow$

$$T = \frac{(\mu + 1) m_1 m_2}{m_1 + m_2} g \qquad (10)$$

If the horizontal surface is *frictionless* ($\mu = 0$), relations (9) and (10) are written:

$$a = \frac{m_2}{m_1 + m_2} g , \quad T = \frac{m_1 m_2}{m_1 + m_2} g .$$

12. *A box of mass m is placed on a cart of mass M, as in the figure. The cart may move on a frictionless horizontal track. The coefficient of static friction between m and M is μ. Find the maximum horizontal force F that we can exert on the cart, in order for the box to move with the cart without sliding on it.*

Solution: We draw the forces separately for each body. We call f the static friction between the two bodies, N the normal force exerted on each body by the other, and R the vertical reaction of the ground on the cart. Note that the direction of f on each body is such as to prevent sliding between these bodies when the exerted force F is directed to the right.

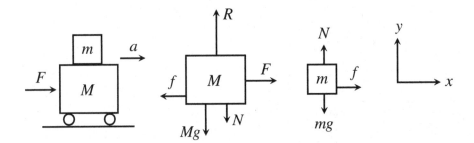

Fig. Problem 12

If no sliding occurs, the two bodies move with a common acceleration, a. We apply Newton's law in the x- and y-directions, for each body separately:

$$F - f = Ma \quad \Rightarrow \quad F = Ma + f \qquad (1)$$

$$R - Mg - N = 0 \quad \Rightarrow \quad R = Mg + N \qquad (2)$$

$$f = ma \tag{3}$$

$$N - mg = 0 \quad \Rightarrow \quad N = mg \tag{4}$$

From (2) and (4) $\Rightarrow$

$$R = (M + m)g.$$

From (1) and (3) $\Rightarrow$

$$F = (M + m)a \tag{5}$$

[Equation (5) is simply Newton's law for the system $(M+m)$]. Now,

$$f \leq f_{\text{max}} = \mu N \overset{(3),(4)}{\Rightarrow} ma \leq \mu mg \quad \Rightarrow \quad a \leq \mu g \tag{6}$$

Equation (6) gives the allowed values of the acceleration in order that the masses move together, with no sliding on each other. From (5) and (6) we find the allowed values of the external force F:

$$F \leq \mu(M + m)g \quad \Leftrightarrow \quad F_{\text{max}} = \mu(M + m)g .$$

13. *A cart, moving on a horizontal track, pushes a box, as shown in the figure. Find the values of the acceleration of the cart in order that the box will not fall to the ground. The coefficient of static friction between the box and the cart is μ.*

Solution: Let a be the common acceleration of the box and the cart. We call m the mass of the box, f the static friction between the box and the cart, and N the normal force on the box by the cart.

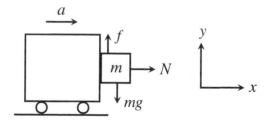

Fig. Problem 13

By Newton's law for the box: $\Sigma \vec{F} = m\vec{a} \Leftrightarrow \Sigma F_x = ma_x, \Sigma F_y = ma_y$, we have:

$$N = ma \tag{1}$$

$$f - mg = 0 \quad \Rightarrow \quad f = mg \qquad (2)$$

But,

$$f \leq f_{max} = \mu N \overset{(1),(2)}{\Rightarrow} mg \leq \mu ma \quad \Rightarrow$$

$$a \geq \frac{g}{\mu} \quad \Leftrightarrow \quad a_{min} = \frac{g}{\mu}.$$

Notice that $a_{min} \to \infty$ when $\mu \to 0$. What does this practically mean?

14. *A box of mass m is placed on a frictionless inclined plane of angle θ, as shown in the figure. The incline is able to move with acceleration of variable magnitude, directed to the right. (a) Find the value of the acceleration of the incline for which the box does not slide on the plane. What is the normal force on the box by the plane in this case? (b) Find the values of the acceleration for which the box moves upward or downward on the incline. (c) For a given value of the acceleration, generally different from that found in part (a), and by assuming now that the plane is* not *frictionless, find the values of the coefficient of static friction μ between the box and the inclined plane in order that the box will not slide on the plane. [Note carefully that in parts (a) and (b) there is no friction between the box and the plane.]*

Solution: We consider an *inertial* reference frame *xy*, which does not accelerate relative to the ground. The *x*-axis is parallel to the incline while the *y*-axis is normal to it. All vectors will be resolved relative to that system of axes. (A system of axes moving with the incline wouldn't be a proper choice since, in view of its acceleration with respect to the ground, it would *not* constitute an inertial frame; hence, we wouldn't be allowed to apply Newton's laws in that frame.)

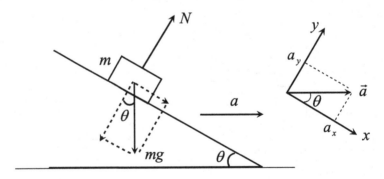

Fig. Problem 14

(a) In order for *m* to not slide on the incline, the acceleration of *m* relative to the *xy*-frame must be equal to the acceleration $\vec{a}$ of the plane. Thus, by Newton's law for *m* we have:

$$\Sigma F_x = ma_x \Rightarrow mg \sin\theta = ma\cos\theta \Rightarrow a = g\tan\theta \tag{1}$$

$$\Sigma F_y = ma_y \Rightarrow N - mg\cos\theta = ma\sin\theta \Rightarrow$$

$$N = m(g\cos\theta + a\sin\theta) \overset{(1)}{=} mg(\cos\theta + \sin\theta\tan\theta).$$

(b) In order for m to slide on the incline, its acceleration $\vec{a}'$ must differ from the acceleration $\vec{a}$ of the plane. Here we only need to consider the x-component of the acceleration (the one parallel to the plane):

$$\Sigma F_x = ma'_x \Rightarrow mg\sin\theta = ma'_x \Rightarrow a'_x = g\sin\theta.$$

The body will slide *downward* if $a_x' > a_x$ or

$$g\sin\theta > a\cos\theta \Rightarrow a < g\tan\theta \quad \text{(downward)} \tag{2}$$

while it will slide *upward* if $a_x' < a_x$ or

$$g\sin\theta < a\cos\theta \Rightarrow a > g\tan\theta \quad \text{(upward)} \tag{3}$$

(c) Conditions (2) and (3) describe the tendency of m to move downward or upward, respectively, relative to the incline when no static friction is present. Hence, friction is necessary in order that m doesn't slide on the plane, for accelerations a of the incline that are different from that given by relation (1). Obviously, the friction f will be directed upward or downward along the plane, depending on whether the acceleration of the latter obeys (2) or (3), respectively. Taking cases (2) and (3) separately, as well as taking into account that

$$f \leq f\text{max} = \mu N,$$

show that

$$\mu \geq \frac{|g\sin\theta - a\cos\theta|}{g\cos\theta + a\sin\theta}.$$

15. *A satellite moves in a circular orbit about the Earth. Inside the satellite an observer is studying the motion of an object of mass m. (a) Explain why a use of*

Newton's laws would lead this observer to physically incorrect conclusions. (b)
As an example, show that, according to this observer, the object m is weightless!

Solution: The satellite is in *free fall* since it moves under the sole action of gravity, with acceleration equal to the acceleration of gravity, $\vec{g}$. (Don't be deceived by the fact that it moves circularly rather than falling straight down to the Earth! The vector $\vec{g}$ is always directed toward the Earth and plays here the role of centripetal acceleration. The satellite would fall straight down hadn't it been given an initial velocity perpendicular to the local direction of the radius of the Earth.) The accelerating satellite is an example of a *non-inertial frame of reference.* An observer moving with the satellite (non-inertial observer) will come to incorrect conclusions if she tries to interpret her measurements physically by using Newton's laws. To understand this, we need to say a few things regarding measurements in non-inertial frames, in general.

We consider two observers: an inertial one, O, at rest on the surface of the Earth, and a non-inertial one, O', moving with an acceleration $\vec{A}$ relative to O. Both observers study the motion of a particle of mass m and record respective accelerations $\vec{a}$ and $\vec{a}'$. By recalling what was said in Sect. 2.8 regarding relative acceleration, it is not hard to see that

$$\vec{a}' = \vec{a} - \vec{A} \tag{1}$$

Since O is an inertial observer, he may use Newton's law to relate the measured acceleration $\vec{a}$ with the total force $\vec{F}$ acting on the particle, which force is due to all interactions in which this particle participates. We thus have:

$$\vec{F} = m\vec{a} \overset{(1)}{\Rightarrow} \vec{F} = m(\vec{a}' + \vec{A}) \Rightarrow m\vec{a}' = \vec{F} - m\vec{A} \equiv \vec{F}' \tag{2}$$

The vector $\vec{a}'$ on the left-hand side of (2) is the acceleration of m as measured by observer O'. We notice that, if observer O' chooses to use Newton's law in her *non-inertial* frame of reference, she will come to the conclusion that, in addition to the actual force $\vec{F}$ (which has the *same* value for all *inertial* observers; cf. Sect. 3.2) the particle is subject to another, *fictitious* force ("pseudo-force") equal to $-m\vec{A}$, which, of course, does not result from any real interactions but is simply an effect related to the acceleration of O'. Thus, if observer O' insists on using Newton's law after measuring the acceleration $\vec{a}'$, she will make a *wrong* evaluation, $\vec{F}'$, of the total force on m, different from the real force $\vec{F}$ evaluated by the inertial observer O.

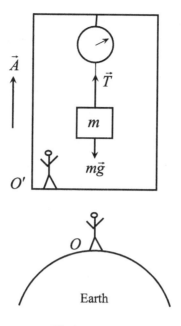

Fig. Problem 15

As an example, assume that the observer O' is in a chamber moving with acceleration $\vec{A}$ relative to the Earth (thus, relative to the inertial observer O who is at rest on the surface of the Earth). Observer O' wants to determine the weight of an object of mass m by using a scale hanging from the ceiling of the chamber, as shown in the figure. The true total force on the body is $\vec{F} = m\vec{g} + \vec{T}$, where $\vec{T}$ is the tension of the string from which the body is suspended. The magnitude T of $\vec{T}$ corresponds to the indication of the scale. The acceleration of m with respect to the inertial observer O is $\vec{A}$. Thus, Newton's law for this observer is written:

$$\vec{F} = m\vec{g} + \vec{T} = m\vec{A} \tag{3}$$

On the other hand, according to the non-inertial observer O' [with respect to whom Newton's law is expressed in the form (2)] the total force on the object m is

$$\vec{F}' = \vec{F} - m\vec{A} \overset{(3)}{\Rightarrow} \vec{F}' = 0.$$

Thus, relation (2) yields $\vec{a}' = 0$, which was to be expected given that m does not accelerate with respect to O'. It is seen from (3) that the scale of O' does not show the actual weight of the body but an *apparent* weight equal to $T = |\vec{T}|$, where

$$\vec{T} = m(\vec{A} - \vec{g}) \tag{4}$$

In particular, if the chamber is a satellite in free fall, then $\vec{A} = \vec{g}$ and it follows from (4) that $\vec{T} = 0$. That is, for an observer O' inside the satellite, the object is *weightless!*

16. *A pendulum of length l and mass m is made to rotate around the vertical with constant angular velocity ω (see figure). Find (a) the angle θ of the string with the vertical, (b) the period T of the motion as a function of θ, and (c) the tension F of the string. (This arrangement is called a **conical pendulum**, since the string describes the surface of an imaginary cone.)*

Solution: The bob m of the pendulum performs uniform circular motion of radius $r = l\sin\theta$ and angular velocity ω, on a horizontal plane that intersects the vertical axis at a distance $l\cos\theta$ below the point of suspension of the pendulum.

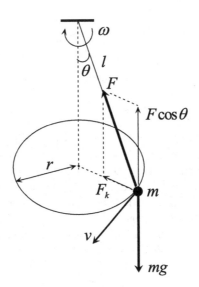

Fig. Problem 16

The forces on m are its weight, $m\vec{g}$, and the tension $\vec{F}$ of the string. Their resultant must yield the centripetal force F_k needed for the uniform circular motion of m. We resolve $\vec{F}$ into a horizontal and a vertical component. Since there is no vertical acceleration, the vertical component $F\cos\theta$ must be equal in magnitude to the weight mg. The only remaining force is the horizontal component $F\sin\theta$, which is exactly the needed centripetal force F_k. We write Newton's law separately for the horizontal and the vertical motion of m:

$$F_k = m\frac{v^2}{r} = mr\omega^2 \quad (\text{since } v = r\omega) \quad \Rightarrow \quad F\sin\theta = mr\omega^2 \tag{1}$$

$$F\cos\theta - mg = 0 \quad \Rightarrow \quad F\cos\theta = mg \tag{2}$$

Dividing (1) by (2) $\Rightarrow \tan\theta = \frac{r\omega^2}{g} = \frac{l\omega^2\sin\theta}{g} \Rightarrow$

$$\cos\theta = \frac{g}{l\omega^2} \tag{3}$$

We notice that θ increases with ω. To find the period T of the circular motion, we write:

$$\omega = \frac{2\pi}{T} = \left(\frac{g}{l\cos\theta}\right)^{1/2} \Rightarrow T = 2\pi\left(\frac{l\cos\theta}{g}\right)^{1/2}.$$

From (2) and (3) we find the tension of the string:

$$F = ml\omega^2.$$

We note the following:

1. As seen from (3), in the limit $\omega \to \infty$ we have $\cos\theta \to 0$ and $\theta \to \pi/2$. Also, for a given l, the angular velocity ω cannot be less than $(g/l)^{1/2}$ (explain).
2. For a given value of θ, the period T is independent of the mass of the pendulum.
3. For a given ω, the tension F does not depend on the value of g (i.e., is independent of gravity).

17. *You may have noticed that railroad tracks and high-speed roads are banked at curves, especially at points of high curvature. This is done in order for a vehicle to be able to make the turn without relying on friction. If θ is the angle of banking of a road at a point where the radius of curvature of the path is r, find the safest speed v with which a car must move at that point if the road is frictionless there. (Assume that v is constant during the turn.)*

Solution: In the figure we see a cross-section of the road and the car. The velocity $\vec{v}$ of the car, of constant magnitude v, is normal to the page (its direction does not matter in the present analysis).

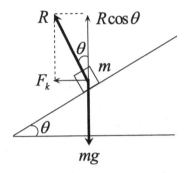

Fig. Problem 17

The forces on the car are its weight, $m\vec{g}$, and the normal reaction $\vec{R}$ from the road (there is no friction). Their resultant must yield the centripetal force F_k needed for the turn. The force F_k is the horizontal component of $\vec{R}$, while the vertical component is balanced by the weight mg (since there is no vertical acceleration). By Newton's law in the horizontal and in the vertical direction, we have:

$$F_k = m\frac{v^2}{r} \quad \Rightarrow \quad R\sin\theta = m\frac{v^2}{r} \tag{1}$$

$$R\cos\theta - mg = 0 \quad \Rightarrow \quad R\cos\theta = mg \tag{2}$$

Dividing (1) by (2), we find:

$$\tan\theta = \frac{v^2}{rg} \quad \Rightarrow \quad v = \sqrt{rg\tan\theta}\,.$$

In practice, of course, the car may move at different speeds with the aid of sidewise and/or frictional forces exerted by the road on the tires.

18. *A train moves uniformly on a curvilinear track that, locally, has a radius of curvature r. A pendulum suspended from the ceiling of the train makes an angle θ with the vertical. Find the local speed v of the train (Assume that v is constant.).*

Solution: The bob of the pendulum, of mass m, moves with the velocity $\vec{v}$ of the train, which is normal to the page in the figure (it doesn't matter whether it is inward or outward).

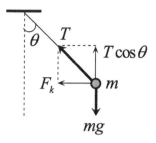

Fig. Problem 18

The forces on m are its weight, $m\vec{g}$, and the tension $\vec{T}$ of the string. Their resultant must yield the required centripetal force F_k in order for m to follow the curvilinear path of the train. The force F_k is the horizontal component of $\vec{T}$, while the vertical component is balanced by the weight mg (there is no vertical acceleration). By Newton's law in the horizontal and in the vertical direction, we have:

$$F_k = m\frac{v^2}{r} \quad \Rightarrow \quad T\sin\theta = m\frac{v^2}{r} \tag{1}$$

$$T\cos\theta - mg = 0 \quad \Rightarrow \quad T\cos\theta = mg \tag{2}$$

Dividing (1) by (2), we get:

$$\tan\theta = \frac{v^2}{rg} \quad \Rightarrow \quad v = \sqrt{rg\tan\theta}\,.$$

19. *According to Newton's Law of Gravity, two masses m_1 and m_2, a distance r apart, attract each other with a force*

$$F = G\frac{m_1 m_2}{r^2}$$

where G is a constant. (For a rigid body, the distance r is measured from the center of mass of the body.) We call M the mass and R the radius of the Earth. (a) Find an expression for the acceleration g of gravity near the surface of the Earth. (b) A satellite moves in a circular orbit about the Earth, with constant speed and at a height h above the surface of the Earth. Find the speed v of the satellite, as well as the period T of the circular motion.

Solution:

(a) Consider a body of mass m located near the surface of the Earth, thus at a distance $r = R$ from its center. The Earth attracts the body with a force equal to the weight of the body:

$$F = G\frac{mM}{R^2} \tag{1}$$

Now, if g is the acceleration of gravity near the surface of the Earth, and if the body is only subject to the gravitational interaction, then, by Newton's second law,

$$F = mg \tag{2}$$

By comparing (1) and (2), we have:

$$g = G\frac{M}{R^2}.$$

Note that g does not depend on m. This means that, near the surface of the Earth, all bodies move under the action of gravity with a common acceleration g (provided, of course, that they are not subject to any additional, non-gravitational interactions).

(b) The radius of the circular path of the satellite is $r = R + h$. Since the speed v of the satellite is constant, the satellite is only subject to a centripetal force equal to the gravitational attraction from the Earth. Hence, if m is the mass of the satellite,

$$m\frac{v^2}{r} = G\frac{mM}{r^2} \quad \Rightarrow \quad v = \sqrt{\frac{GM}{r}} = \sqrt{\frac{GM}{R+h}} \tag{3}$$

Note again that the result is independent of m. We also have:

$$v = \omega r = \frac{2\pi r}{T} \quad \Rightarrow \quad T = \frac{2\pi r}{v} \overset{(3)}{\Rightarrow} T = \frac{2\pi(R+h)^{3/2}}{\sqrt{GM}}.$$

20. *A particle moves on the xy-plane under the action of a force $\vec{F} = \vec{v} \times \vec{A}$, where $\vec{v}$ is the velocity of the particle and where $\vec{A}$ is a constant vector parallel to the z-axis. (a) Show that $\vec{F}$ is a vector belonging to the xy-plane and determine the direction of $\vec{F}$ relative to the trajectory of the particle. (b) Show that $\vec{v}$ and $\vec{F}$ are vectors of constant magnitude. (c) Show that the particle executes uniform circular motion.*

Solution:

(a) Let $\vec{v} = v_x \hat{u}_x + v_y \hat{u}_y$ and $\vec{A} = A\hat{u}_z$, where $A = |\vec{A}|$ (we chose $\vec{A}$ to be in the positive direction of the z-axis). Then,

$$\vec{F} = \vec{v} \times \vec{A} = (v_x \hat{u}_x + v_y \hat{u}_y) \times (A\hat{u}_z) = A(v_y \hat{u}_x - v_x \hat{u}_y)$$

which is a vector in the xy-plane. Moreover, by the definition of the vector product, $\vec{F}$ is normal to the velocity $\vec{v}$, hence normal to the trajectory. This can also be seen as follows:

$$\vec{F} \cdot \vec{v} = A(v_y \hat{u}_x - v_x \hat{u}_y) \cdot (v_x \hat{u}_x + v_y \hat{u}_y) = A(v_y v_x - v_x v_y) = 0 .$$

(b) Since the total force $\vec{F}$ is normal to the velocity, the speed $|\vec{v}| = v$ is constant (only the direction of motion changes). Furthermore,

$$|\vec{F}| \equiv F = |\vec{v}||\vec{A}| \sin \frac{\pi}{2} = vA = \text{constant} .$$

(c) Let ρ be the radius of curvature at an arbitrary point of the trajectory, and let m be the mass of the particle. Since $\vec{F}$ is centripetal at all points of the path,

$$F = m\frac{v^2}{\rho} \quad \Rightarrow \quad \rho = \frac{mv^2}{F} = \text{constant}$$

(since both v and F are constant). We thus have a uniform motion on a plane, along a path with constant radius of curvature. What kind of motion is this?

21. *A particle of mass m is moving on the xy-plane. Its coordinates are given as functions of time by the equations: $x = A \cos \omega t$, $y = A \sin \omega t$, where A, ω are positive constants.*

 (a) *Find the equation of the trajectory of the particle, in the form $F(x,y) = const$.*

 (b) *Find the velocity $\vec{v} \equiv (v_x, v_y)$ and the acceleration $\vec{a} \equiv (a_x, a_y)$ of the particle as functions of time and show that these vectors are mutually perpendicular at every point of the trajectory. What do you conclude about the speed of the particle? Verify your conclusion by calculating this speed directly.*

 (c) *Find the tangential and the normal acceleration as functions of time.*

(d) *Find the radius of curvature, ρ, as a function of time and show that your result is consistent with that of part (a). Characterize the type of motion of the particle.*

(e) *Show that the particle is subject to a total force of constant magnitude. How is this force oriented relative to the trajectory?*

(f) *Show that the angular momentum of the particle, with respect to the origin O of the coordinate system (x, y), is constant in time. What do you conclude regarding the total torque on the particle about O?*

Solution:

(a) $\cos \omega t = \frac{x}{A}$, $\sin \omega t = \frac{y}{A}$. Then,

$$\cos^2 \omega t + \sin^2 \omega t = 1 \quad \Rightarrow \quad \frac{x^2}{A^2} + \frac{y^2}{A^2} = 1 \quad \Rightarrow \quad x^2 + y^2 = A^2 .$$

The trajectory is a circle of radius A, centered at the origin O $(x = y = 0)$ of the coordinate system.

(b) We have:

$$v_x = \frac{dx}{dt} = -\omega A \sin \omega t , \quad v_y = \frac{dy}{dt} = \omega A \cos \omega t$$
$$a_x = \frac{dv_x}{dt} = -\omega^2 A \cos \omega t , \quad a_y = \frac{dv_y}{dt} = -\omega^2 A \sin \omega t$$

Hence, $\vec{v} \cdot \vec{a} = v_x a_x + v_y a_y = \omega^3 A^2 (\sin \omega t \cos \omega t - \cos \omega t \sin \omega t) = 0.$

The acceleration $\vec{a}$ is normal to the velocity $\vec{v}$; thus $|\vec{v}| = $ constant. Indeed:

$$|\vec{v}|^2 = v_x^2 + v_y^2 = \omega^2 A^2 \quad \Rightarrow \quad |\vec{v}| \equiv v = \omega A = \text{constant} \tag{1}$$

(c) $a_T = \frac{dv}{dt} = 0$, since $v = |\vec{v}| = $ constant . To find a_N we work as follows:

$$|\vec{a}|^2 = a_T^2 + a_N^2 = a_x^2 + a_y^2 \Rightarrow 0 + a_N^2 = \omega^4 A^2 \Rightarrow a_N = \omega^2 A \tag{2}$$

(d) $a_N = \frac{v^2}{\rho} \Rightarrow \rho = \frac{v^2}{a_N} \overset{(1),(2)}{\Rightarrow} \rho = A = $ constant.

The motion is uniform circular, of radius A, centered at O.

(e) Since the motion is uniform, the total force is purely centripetal:

$$F = m a_N \overset{(2)}{\Rightarrow} F = m \omega^2 A = \text{constant} .$$

(f) The position vector of the particle, relative to O, is $\vec{r} = x\hat{u}_x + y\hat{u}_y$. The angular momentum with respect to O is

$$\vec{L} = m(\vec{r} \times \vec{v}) = m(x\hat{u}_x + y\hat{u}_y) \times (v_x\hat{u}_x + v_y\hat{u}_y) \quad \Rightarrow$$

$$\vec{L} = m(xv_y - yv_x)\hat{u}_z = m\omega A^2\hat{u}_z .$$

Alternatively, by relation (3.32), $\vec{L} = L\hat{u}_z$ where $L = |\vec{L}| = mA^2\omega$. We notice that $\vec{L} = $ constant. Thus there is no torque about O, given that, in general, $\vec{T} = d\vec{L}/dt$.

Problems for Chaps. 4–6

22. *A particle of mass m is moving on a circle of radius R. We call L the magnitude of the angular momentum of the particle relative to the center O of the circle. (a) Show that the kinetic energy of the particle is*

$$E_k = \frac{L^2}{2mR^2} = \frac{L^2}{2I}$$

where $I = mR^2$ is the moment of inertia of m with respect to the axis of rotation (this axis is normal to the plane of the circle and passes through the center of the circle). (b) If the circular motion is uniform, show that the magnitude of the total force on the particle is

$$F = \frac{L^2}{mR^3} .$$

Solution:

(a) We know that

$$E_k = \frac{1}{2}mv^2 = \frac{p^2}{2m} \quad \text{where} \quad p = mv .$$

Moreover, for circular motion [see Eq. (3.32)],

$$L = mRv = Rp \quad \Rightarrow \quad p = \frac{L}{R} .$$

Thus,

$$E_k = \frac{1}{2m}\left(\frac{L}{R}\right)^2 = \frac{L^2}{2mR^2} = \frac{L^2}{2I} .$$

(b) If the motion is uniform, the total force is purely centripetal:

$$F = m\frac{v^2}{R} = \frac{m}{R}\left(\frac{L}{mR}\right)^2 = \frac{L^2}{mR^3} \,.$$

23. *An electrically charged particle of charge q, moving with velocity $\vec{v}$ inside a magnetic field of strength $\vec{B}$, is subject to a magnetic force $\vec{F} = q(\vec{v} \times \vec{B})$. (a) Show that $\vec{F}$ does not produce any work. (b) If no other forces act on the particle, what do you conclude regarding the particle's kinetic energy?*

Solution:

(a) The work of $\vec{F}$ from a point a to a point b is

$$W = \int_a^b \vec{F} \cdot d\vec{r} = \int_a^b \vec{F} \cdot \frac{d\vec{r}}{dt}\, dt = \int_a^b (\vec{F} \cdot \vec{v})\, dt = 0$$

since

$$\vec{F} \cdot \vec{v} = q(\vec{v} \times \vec{B}) \cdot \vec{v} = 0 \quad \text{(explain why!)}\,.$$

(b) By the work-energy theorem, $W = \Delta E_k = 0 \Rightarrow E_k = $ constant. Alternatively, since the total force $\vec{F}$ is normal to the velocity $\vec{v}$, the speed v is constant. Hence, $E_k = mv^2/2 = $ constant.

24. *A stone of weight 20 N falls from a height $h = 10$ m and sinks by $h' = 0.5$ m into the ground. Find the average force, f, from the ground to the stone as the latter sinks.*

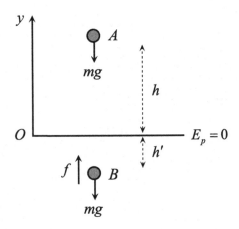

Fig. Problem 24

Solution: The ground is at height $y = 0$. The velocity of the stone at both the initial point A and the final point B (see figure) is zero. The stone is subject to the conservative force mg and the non-conservative force f. The change of the total mechanical energy $(E_k + E_p)$ of the stone from A to B is equal to the work of f :

$$\Delta(E_k + E_p) = (E_k + E_p)_B - (E_k + E_p)_A = W_f \quad \Rightarrow$$
$$(0 - mgh') - (0 + mgh) = -fh' \quad \Rightarrow$$
$$f = \frac{mg(h + h')}{h'} = 420\,N\,.$$

(The potential energy at B is negative since B is located below the reference level $y = 0$. The work of f is negative since f opposes the motion of the stone.)

25. *Show that the tension of the string of an oscillating pendulum is $T = mg\,(3\cos\theta - 2\cos\theta_0)$, where θ is the instantaneous angle of the string with the vertical while θ_0 is the angular amplitude of oscillation.*

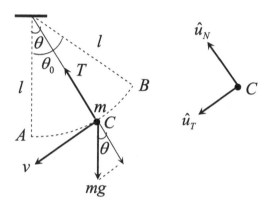

Fig. Problem 25

Solution: We call l the length of the string. The forces on m (see figure) are its weight, mg, and the tension T of the string. We resolve the motion into two mutually perpendicular directions, namely, a tangential one, parallel to the velocity, and a centripetal one, parallel to the string. We want to apply Newton's law in each direction. To this end, we introduce the unit vectors $\hat{u}_T$ and $\hat{u}_N$ (drawn separately in the figure) at the instantaneous position C of m. (Recall that $\hat{u}_T$ is tangent to the path while $\hat{u}_N$ is normal to it and directed "inward". The direction of $\hat{u}_T$ was chosen arbitrarily.) We call $\vec{F}$ the total force on m and we let $\vec{a}$ be the acceleration of m :

$$\vec{F} = m\vec{g} + \vec{T} = m\vec{a} \Rightarrow$$
$$(mg\sin\theta\,\hat{u}_T - mg\cos\theta\,\hat{u}_N) + T\hat{u}_N = m\left(\frac{dv}{dt}\hat{u}_T + \frac{v^2}{l}\hat{u}_N\right) \quad \Rightarrow$$

$$mg\sin\theta\,\hat{u}_T + (T - mg\cos\theta)\hat{u}_N = m\frac{dv}{dt}\hat{u}_T + m\frac{v^2}{l}\hat{u}_N \tag{1}$$

By equating the coefficients of $\hat{u}_N$ and $\hat{u}_T$ on the two sides of (1) $\Rightarrow$

$$T - mg\cos\theta = m\frac{v^2}{l} \Rightarrow T = m\left(g\cos\theta + \frac{v^2}{l}\right) \tag{2}$$

$$mg\sin\theta = m\frac{dv}{dt} \Rightarrow \frac{dv}{dt} = g\sin\theta \tag{3}$$

Relation (3) is a differential equation for the algebraic value v of the velocity. Its solution is not an easy task, thus we try to solve the equation by using conservation of energy. Let E_p be the gravitational potential energy of m. As we know, the change of the total mechanical energy $(E_k + E_p)$ is equal to the work of the forces that do not contribute to the potential energy: $\Delta(E_k + E_p) = W'$. Here, W' is the work of the tension T. Given that T is always normal to the velocity, its work vanishes: $W' = 0$. Hence, $\Delta(E_k + E_p) = 0 \Leftrightarrow E_k + E_p = $ constant. In particular,

$$\left(E_k + E_p\right)_c = \left(E_k + E_p\right)_B \tag{4}$$

We choose $E_p = 0$ at the lowest point A (where $\theta = 0$). Thus, at an arbitrary point C where the angle is θ,

$$E_p = mg(l - l\cos\theta) = mgl(1 - \cos\theta)$$

[where $(l - l\cos\theta)$ is the *vertical* distance of C from A]. At the highest point B of the path we have $\theta = \theta_0$ and $v = 0$. Equation (4) then yields:

$$\frac{1}{2}mv^2 + mgl(1 - \cos\theta) = 0 + mgl(1 - \cos\theta_0) \quad \Rightarrow$$

$$v^2 = 2gl(\cos\theta - \cos\theta_0) \tag{5}$$

Substituting (5) into (2), we find:

$$T = mg(3\cos\theta - 2\cos\theta_0).$$

Note that the result is independent of l.

Exercise: Show that the expression (5) for v satisfies the differential equation (3). [*Hint*: Differentiate (5) with respect to t and use the fact that $v = -l d\theta/dt$ (the minus sign is related to the fact that v is negative when the angle θ increases).]

26. *A bead of mass m is at rest on the highest point A of a frictionless spherical surface of radius r (see figure). We give the bead a slight push, just to make it*

slide on the surface. (a) Find the normal reaction R of the surface on the bead, as a function of the angle θ. (b) Find the angle θ at which the bead will detach from the sphere.

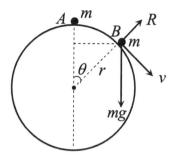

Fig. Problem 26

Solution: Let B be an arbitrary position of m on the sphere, corresponding to an angle $θ$. We work as in Problem 25. We introduce the unit vectors $\hat{u}_T$ and $\hat{u}_N$ at B, tangential and normal, respectively, to the surface (draw them). We then apply Newton's law and take its components in the directions of these unit vectors. Newton's law for the centripetal force is written:

$$mg\cos θ - R = m\frac{v^2}{r} \quad \Rightarrow \quad R = m(g\cos θ - \frac{v^2}{r}) \qquad (1)$$

The tangential component of the law leads to a differential equation for the algebraic value v of the velocity. An alternative approach is by conservation of energy. Write: $Δ(E_k + E_p) = W'$, where E_p is the gravitational potential energy and W' is the work of the normal reaction R (the force that does not contribute to the potential energy). Here, $W' = 0$, since R is always normal to the velocity. Thus, $Δ(E_k + E_p) = 0 \Leftrightarrow E_k + E_p = $ constant. In particular,

$$\left(E_k + E_p\right)_A = \left(E_k + E_p\right)_B \qquad (2)$$

We choose $E_p = 0$ at the equilibrium point A (where $θ = 0$). Thus, at the arbitrary position B corresponding to an angle $θ$,

$$E_p = -mg(r - r\cos θ) = -mgr(1 - \cos θ).$$

[The point B is at a *vertical* distance $(r - r\cos θ)$ *below* the reference point A. This also explains the negative value of E_p at B.] At the highest point A we have $θ = 0$ and $v = 0$. Equation (2) then yields:

$$0 + 0 = \frac{1}{2}mv^2 - mgr(1 - \cos θ) \quad \Rightarrow$$

$$v^2 = 2gr(1 - \cos\theta) \tag{3}$$

Substituting (3) into (1), we find:

$$R = mg(3\cos\theta - 2).$$

For the bead to remain in contact with the sphere, we must have $R \geq 0 \Rightarrow \cos\theta \geq 2/3$. Thus the bead will detach from the sphere when $\theta = \text{arc} \cos(2/3)$.

27. *In an ice-hockey game the player hits the puck giving it an initial velocity $v_0 = 20$ m/s. The puck slides on the ice, moving rectilinearly, and travels a distance $\Delta x = 120$ m until it stops. Find the coefficient of kinetic friction, μ, between the puck and the ice. ($g = 9.8$ m/s^2)*

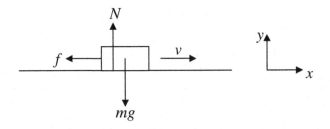

Fig. Problem 27

Solution: The forces on the puck are its weight, mg, the normal force N from the ice, and the kinetic friction f (see figure). We apply Newton's law in the x and y directions:

$$\Sigma F_x = ma_x \quad \Rightarrow \quad -f = ma \tag{1}$$

$$\Sigma F_y = ma_y \quad \Rightarrow \quad N - mg = 0 \quad \Rightarrow \quad N = mg \tag{2}$$

where $a\ (= a_x)$ is the algebraic value of the acceleration of the puck. Relations (1) and (2), in combination with $f = \mu N$, yield:

$$ma = -\mu N = -\mu mg \quad \Rightarrow \quad a = -\mu g \tag{3}$$

Notice that $a < 0$ while $v > 0$, so that $va < 0$. This means that the motion is retarded (cf. Sect. 2.4) and indeed uniformly, since the acceleration a is constant. We may thus use the relation

$$v^2 = v_0^2 + 2a(x - x_0) = v_0^2 + 2a\Delta x$$

where Δx is the distance traveled by m. For the given distance Δx, at the end of which the object stops, the final velocity is $v = 0$. Hence,

$$0 = v_0^2 + 2a\Delta x \quad \Rightarrow \quad a = -\frac{v_0^2}{2\Delta x} \tag{4}$$

By comparing (3) with (4), we find:

$$\mu = \frac{v_0^2}{2g\Delta x} \tag{5}$$

Alternatively, we can make use of the work-energy theorem. The forces N and mg do not produce work since they are normal to the velocity. Thus the total work on m is done by the friction f alone and is equal to the change of the kinetic energy of m:

$$\Delta E_k = W_f \quad \Rightarrow \quad 0 - \frac{1}{2}mv_0^2 = -f\,\Delta x.$$

(Explain the negative sign in the expression for W_f.) We have:

$$\frac{1}{2}mv_0^2 = f\,\Delta x = \mu N\,\Delta x = \mu mg\,\Delta x \quad \Rightarrow \quad \mu = \frac{v_0^2}{2g\Delta x},$$

which is the same as (5). Substituting for the given values, we find: $\mu = 0.17$.

28. *A cube is let to slide down form the top of an inclined plane of length l and angle of inclination θ, which plane rests on a horizontal table (see figure). There is no friction between the cube and the incline, while the coefficient of kinetic friction between the cube and the table is μ. How far from the base of the incline will the cube go before it stops?*

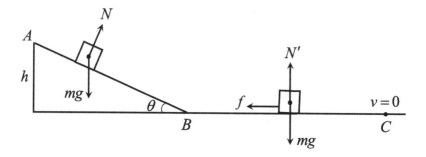

Fig. Problem 28

Solution: Along the path AB, of length $AB = l$, the cube is subject to the conservative force mg and the normal reaction N, which produces no work (why?). Thus,

$$\Delta\left(E_k + E_p\right) = W_N = 0 \quad \Rightarrow \quad E_k + E_p = \text{constant}.$$

In particular,

$$\left(E_k + E_p\right)_A = \left(E_k + E_p\right)_B.$$

Taking $E_p = 0$ at the base of the inclined plane, and using the fact that $v_A = 0$, we have:

$$0 + mgh = \frac{1}{2}mv_B^2 + 0 \quad \Rightarrow \quad v_B^2 = 2gh = 2gl\sin\theta \tag{1}$$

Along the path BC (where C is the point at which the cube stops) only the friction f produces work (why?), which work is negative since f opposes the motion of the cube. By the work-energy theorem,

$$\Delta E_k = E_{k,C} - E_{k,B} = W_f \quad \Rightarrow \quad 0 - \frac{1}{2}mv_B^2 = -f(BC) \quad \Rightarrow$$

$$mv_B^2 = 2f(BC) = 2\mu N'(BC) = 2\mu mg(BC) \quad \Rightarrow \quad v_B^2 = 2\mu g(BC) \tag{2}$$

By comparing (1) with (2), we find:

$$BC = \frac{h}{\mu} = \frac{l\sin\theta}{\mu}.$$

29. *The mass m in the figure is initially at a location A at which the two springs have their natural length (they are not deformed). The mass is then displaced a distance l to the right, at point B. There is no friction between m and the horizontal surface. (a) If m is set free at B, find its velocity at the moment when it passes from A. (b) Find the maximum distance to the left of A, traveled by m. Given: m = 4 kg, l = 0.2 m, k = 8 N/m, k' = 5 N/m.*

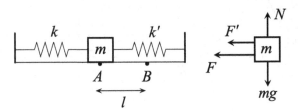

Fig. Problem 29

Solution: The forces on m are its weight, mg, the normal reaction N from the horizontal surface, and the elastic forces F and F' by the springs k and k', respectively. The reaction N does not produce work (why?) while the gravitational potential energy of m is constant, conveniently chosen to be zero. Thus, the total potential energy of m is due to the springs alone and equal to

$$E_P = \frac{1}{2}k(\Delta l)^2 + \frac{1}{2}k'(\Delta l)^2 = \frac{1}{2}(k + k')(\Delta l)^2$$

where Δl is the common deformation of the two springs. Furthermore,

$$\Delta(E_k + E_p) = W_N = 0 \quad \Rightarrow \quad E_k + E_p = \text{constant}.$$

(a) At location B, $\Delta l = l$ and $v_B = 0$, while at A, $\Delta l = 0$. Thus,

$$(E_k + E_p)_B = (E_k + E_p)_A \quad \Rightarrow \quad 0 + \frac{1}{2}(k + k')l^2 = \frac{1}{2}mv_A^2 + 0 \Rightarrow$$

$$v_A = l\sqrt{\frac{k + k'}{m}} = 0.36\,m/s.$$

(b) Assume that m stops momentarily at some point C that is a distance x to the left of A. Then, $v_B = v_C = 0$ and

$$(E_k + E_p)_B = (E_k + E_p)_C \quad \Rightarrow \quad 0 + \frac{1}{2}(k + k')l^2 = 0 + \frac{1}{2}(k + k')x^2 \quad \Rightarrow$$

$$x = l = 0.2\,m.$$

30. *A body of mass m is fired from the surface of the Earth. Find the **escape velocity** of the body, that is, the minimum ejection speed necessary in order for the body to free itself from the gravitational attraction of the Earth and reach "infinity". Given: the mass M and the radius R of the Earth. (The potential energy of m in the gravitational field of the Earth is*

$$E_p(r) = -G\frac{Mm}{r}$$

where G is a constant and where $r \geq R$ is the distance of m from the center of the Earth.)

Solution: Let v_0 be the ejection speed. The total mechanical energy of m at the moment of ejection is

$$E_0 = E_{k,0} + E_p(R) = \frac{1}{2}mv_0^2 - G\frac{Mm}{R}$$

(since $r = R$ on the surface of the Earth). Assume now that m has sufficient energy to escape from the Earth and reach "infinity" ($r = \infty$) with some speed v_∞. The total energy of m at infinity is

$$E_\infty = E_{k,\infty} + E_p(\infty) = \frac{1}{2}mv_\infty^2 + 0 = \frac{1}{2}mv_\infty^2 .$$

By conservation of mechanical energy (if we ignore air resistance),

$$E_0 = E_\infty \quad \Rightarrow \quad \frac{1}{2}mv_0^2 - G\frac{Mm}{R} = \frac{1}{2}mv_\infty^2 \geq 0 \quad \Rightarrow \quad v_0^2 \geq \frac{2GM}{R} \quad \Rightarrow$$

$$v_0 \geq \sqrt{\frac{2GM}{R}} .$$

The minimum ejection speed is

$$v_{0,\min} = \sqrt{\frac{2GM}{R}}$$

and we notice that it is independent of the mass m of the body, as well as of the direction of ejection from the Earth. (Note, however, that the required kinetic energy for escape *does* depend on the mass of the body!)

31. *Two masses m_1 and m_2 are connected by a vertical spring of constant k, as shown in the figure. The upper mass m_1 is at rest at a position A. We press m_1 downward to position B and then let it free. The mass m_1 moves upward to C and stops there momentarily before it starts moving downward again. (a) Show that the vertical distances AB and AC are equal. (b) Find the vertical displacement AB of m_1 in order that the mass m_2 will rise from the ground when m_1 reaches C.*

Solution: At positions A and B the spring is obviously compressed relative to its natural length. We call x_0 the compression of the spring when m_1 is at the equilibrium position A. In order for m_2 to rise from the ground, the spring must be in *extension* when m_1 is at C. Therefore, C must be located above the height y that corresponds to the natural length of the free spring.

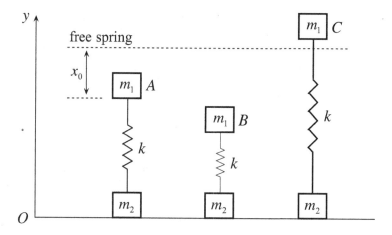

Fig. Problem 31

The equilibrium condition for m_1 at point A is

$$kx_0 = m_1 g \qquad (1)$$

The total mechanical energy of m_1 at an arbitrary position y is

$$E = E_k + E_p = \frac{1}{2}m_1 v^2 + m_1 g y + \frac{1}{2}kx^2$$

where x is the deformation of the spring (compression or extension) relative to its natural length. At positions B and C, $v_B = v_C = 0$. Also, $x_B = x_0 + AB$, $x_C = AC - x_0$. By conservation of mechanical energy, $E_B = E_C \Rightarrow$

$$0 + m_1 g y_B + \frac{1}{2}k(x_0 + AB)^2 = 0 + m_1 g y_C + \frac{1}{2}k(AC - x_0)^2 \Rightarrow$$

$$\frac{1}{2}k\left[(x_0 + AB)^2 - (AC - x_0)^2\right] = m_1 g(y_C - y_B) = m_1 g(BC) \Rightarrow$$

$$\frac{1}{2}k\left[(AB)^2 - (AC)^2\right] + kx_0(AB + AC) = m_1 g(BC)$$

or, given that $AB + AC = BC$,

$$\frac{1}{2}k\left[(AB)^2 - (AC)^2\right] = (m_1 g - kx_0)(BC) = 0$$

due to (1). Thus, finally,

$$AB = AC \qquad (2)$$

In order for the mass m_2 to rise (even slightly), the extended spring must exert on it an upward force at least equal to its weight:

$$kx_C \geq m_2 g \quad \Rightarrow \quad k(AC - x_0) \geq m_2 g \, .$$

Substituting $AC = AB$ and $k\, x_0 = m_1 \, g$, according to (1) and (2), we find:

$$AB \geq \frac{(m_1 + m_2)g}{k} \, .$$

32. *A pump raises water at a height H above a tank, at a rate of α kg/s (mass of water per unit time). The water reaches its final height having acquired a speed v. Find the work supplied per unit time by the pump.*

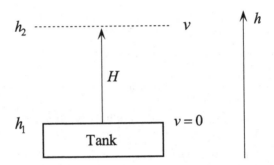

Fig. Problem 32

Solution: The heights h_1 and h_2 (see figure) are measured relative to an arbitrary reference level. Let dm be the mass of water arriving at height h_2 and, within time dt, added to the quantity of water already existing there. The rate at which water is delivered at h_2 is

$$\frac{dm}{dt} = \alpha \tag{1}$$

The work dW supplied by the pump within time dt is equal to the change of the total mechanical energy of dm between the levels h_1 and h_2:

$$dW = \Delta(E_k + E_p) = (E_k + E_p)_2 - (E_k + E_p)_1 = \left[\frac{1}{2}(dm)v^2 + (dm)gh_2\right] - \left[0 + (dm)gh_1\right].$$

That is,

$$dW = \frac{1}{2}(dm)v^2 + (dm)g(h_2 - h_1) = \frac{1}{2}(dm)v^2 + (dm)gH \, .$$

The work supplied per unit time by the pump is, therefore,

$$\frac{dW}{dt} = \frac{1}{2}\frac{dm}{dt}v^2 + \frac{dm}{dt}gH \overset{(1)}{\Rightarrow}$$

$$\frac{dW}{dt} = \frac{1}{2}\alpha v^2 + \alpha gH .$$

33. *A particle performs simple harmonic motion on the x-axis, according to the equation $x = A \cos\omega t$, where $\omega = 2\pi/T$. Determine the time intervals in which the motion is accelerated or retarded.*

Solution: In accordance with the discussion in Sect. 2.4, we need to examine the sign of the product va, where v and a are the algebraic values of the velocity and the acceleration, respectively. We have:

$$v = \frac{dx}{dt} = -\omega A \sin \omega t, \quad a = \frac{dv}{dt} = -\omega^2 A \cos \omega t \quad \Rightarrow$$

$$va = \omega^3 A^2 \sin \omega t \cos \omega t = \frac{\omega^3 A^2}{2} \sin 2\omega t .$$

The motion is *accelerated* when

$$va > 0 \quad \Rightarrow \quad \sin 2\omega t > 0 \quad \Rightarrow \quad 2k\pi < 2\omega t < (2k+1)\pi \quad \Rightarrow$$

$$k\pi < \omega t = \frac{2\pi t}{T} < k\pi + \frac{\pi}{2} \quad \Rightarrow \quad \frac{kT}{2} < t < \frac{(2k+1)T}{4}$$

where $k = 0, 1, 2, \ldots$ Putting $k = 0$ and $k = 1$, we find $0 < t < T/4$ and $T/2 < t < 3T/4$, respectively. (Determine the intervals where the motion is retarded.)

34. *A mass m may be connected to two springs in three different ways, as shown in figure 1. Find the period of oscillation of m in each case. [The masses of the springs are negligible. In case (c) the horizontal surface is frictionless and both springs are extended at the equilibrium position of m.]*

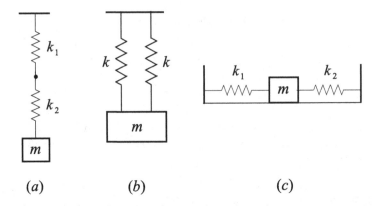

Fig. Problem 34 (1)

Solution: We will follow a process similar to that used in problems with resistors or capacitors in electricity. We will try to replace the two-spring system with an equivalent single spring that, when suffering deformation equal to that of the system, it exerts the same force at its end. We recall (Sect. 5.4) that, when a spring of constant k is deformed (either extended or compressed) by x, it exerts forces of magnitude $F = kx$ at its ends, opposing the deformation. Now, if we connect one end of the spring with a mass m while keeping the other end fixed, the system will execute harmonic oscillations about the equilibrium position of m, with period

$$T = 2\pi \sqrt{\frac{m}{k}} \tag{1}$$

a. *Vertical springs k_1 and k_2 connected in series* (see figure 2)

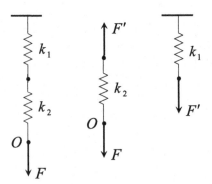

Fig. Problem 34 (2)

Assume that we apply a force F at the end O of the system of springs. By the action-reaction law, the system will exert on us an opposite force of equal magnitude F. Let x_1 and x_2 be the individual extensions of k_1 and k_2, respectively. The total extension of the system will be $x = x_1 + x_2$. We will show that the force F exerted by the system is proportional to the total deformation x of the system, relative to its natural length.

We call F' the internal force from one spring to the other. By Newton's law, since k_2 has negligible mass, the total force on it is zero; thus $F' = F$. Taking into account the individual extensions of the springs, we have:

$$F = k_2 x_2 = k_1 x_1 \quad \Rightarrow \quad x_1 = F/k_1, \quad x_2 = F/k_2.$$

Hence,

$$x = x_1 + x_2 = \frac{F}{k_1} + \frac{F}{k_2} = \left(\frac{1}{k_1} + \frac{1}{k_2}\right) F \tag{2}$$

We define the *equivalent constant* k' by

$$\frac{1}{k'} = \frac{1}{k_1} + \frac{1}{k_2} \quad \Leftrightarrow \quad k' = \frac{k_1 k_2}{k_1 + k_2}.$$

Relation (2) is then written:

$$x = \frac{1}{k'} F \quad \Rightarrow \quad F = k' x.$$

That is, when the system is extended by x, it exerts a force proportional to x, as if it were a single spring of constant k'. If we now suspend a mass m at its end O, the system will oscillate about the equilibrium position of m with period given by (1):

$$T = 2\pi \sqrt{\frac{m}{k'}} = 2\pi \left(\frac{m}{k_1} + \frac{m}{k_2}\right)^{1/2}.$$

In particular, if $k_1 = k_2 = k$, then $k' = k/2$ and $T = 2\pi\sqrt{2m/k}$.

b. *Vertical springs k connected in parallel* (see figure 3)

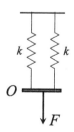

Fig. Problem 34 (3)

Assume that we apply a force F at the end O of the system, causing the system an extension x from its natural length. Obviously, both springs will be extended by x. By the action-reaction law, the system will exert on us an opposite force of the same magnitude F, equal to the resultant of the forces exerted by each spring separately:

$$F = kx + kx = 2kx \equiv k'x \quad \text{where } k' = 2k .$$

If we now suspend a mass m at O, the system will oscillate with period

$$T = 2\pi \sqrt{\frac{m}{k'}} = 2\pi \sqrt{\frac{m}{2k}} .$$

c. *Springs k_1 and k_2 connected to a mass m on a frictionless horizontal surface* (see figure 4)

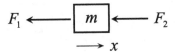

Fig. Problem 34 (4)

Initially, the mass m balances while the springs are extended. Assume now that we displace m by x to the right. The extension of k_1 will thus increase by x, while that of k_2 will decreases by x. The mass m will then be subject to *additional* (unbalanced) forces $F_1 = k_1 x$ and $F_2 = k_2 x$, *both* opposite to the displacement. Given the absence of friction, the total force on m will be

$$F = F_1 + F_2 = (k_1 + k_2)x \equiv k'x , \quad \text{where } k' = k_1 + k_2 .$$

The period of oscillation is

$$T = 2\pi \sqrt{\frac{m}{k'}} = 2\pi \sqrt{\frac{m}{k_1 + k_2}} .$$

35. *An elevator of mass M is suspended from a spring of constant k, as shown in the figure. A box of mass m rests on the floor of the elevator. The system is displaced vertically by a distance A from its equilibrium position and is then let free. Find the values of A for which the box will not separate from the elevator.*

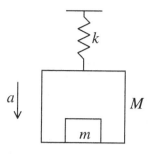

Fig. Problem 35

Solution: We assume that the elevator and the box move together like a single body, with common acceleration a. The situation is thus equivalent to that of a mass $(M + m)$ hanging from a spring k. Such a system executes harmonic oscillations about its equilibrium position, of angular frequency $\omega = [k/(M + m)]^{1/2}$. If y is the instantaneous displacement from the equilibrium position, the corresponding *total* force on the system has magnitude $F = ky$ [see Sect. 5.4, Eq. (5.20)]. Thus, the magnitude of the instantaneous acceleration of the system is

$$a = \frac{F}{M + m} = \frac{ky}{M + m}.$$

Given that the maximum displacement equals the amplitude A of oscillation, the maximum value of the acceleration is

$$a_{\max} = \frac{kA}{M + m} \tag{1}$$

Now, the box may separate from the floor of the elevator only when the system accelerates downward. Thus, if separation is to be avoided, the downward acceleration a of the elevator must never exceed the acceleration g of free fall of the box:

$$a_{\max} \leq g \overset{(1)}{\Rightarrow} A \leq \frac{(M + m)g}{k}.$$

36. *Consider two particles m_1 and m_2 subject only to their mutual interaction (isolated system). Call $\vec{F}_{12}$ the force on m_1 due to m_2, and $\vec{a}_{12}$ the acceleration of m_1 relative to m_2. (Then, $\vec{F}_{21} = -\vec{F}_{12}$ and $\vec{a}_{21} = -\vec{a}_{12}$.) Show that*

$$\vec{F}_{12} = \mu \vec{a}_{12} \tag{1}$$

*where μ is the **reduced mass** of the system:*

$$\mu = \frac{m_1 m_2}{m_1 + m_2} \quad \Leftrightarrow \quad \frac{1}{\mu} = \frac{1}{m_1} + \frac{1}{m_2}.$$

Solution: Let $\vec{a}_1, \vec{a}_2$ be the accelerations of m_1, m_2, relative to an *inertial* frame of reference. By Newton's law, and given the absence of external forces, we have:

$$\vec{a}_1 = \frac{\vec{F}_{12}}{m_1}, \quad \vec{a}_2 = \frac{\vec{F}_{21}}{m_2} = -\frac{\vec{F}_{12}}{m_2}.$$

Then,

$$\vec{a}_{12} = \vec{a}_1 - \vec{a}_2 = \left(\frac{1}{m_1} + \frac{1}{m_2}\right)\vec{F}_{12} = \frac{1}{\mu}\vec{F}_{12} \quad \Rightarrow \tag{1}$$

37. *Show that conservation of kinetic energy is impossible in a completely inelastic (plastic) collision where two bodies stick together.*

Solution: Assume that the masses m_1 and m_2 are moving along the *x*-axis and, just before colliding, their velocities are $\vec{v}_1 = v_1 \hat{u}_x$ and $\vec{v}_2 = v_2 \hat{u}_x$, while right after the collision the composite mass $(m_1 + m_2)$ has velocity $\vec{V} = V\hat{u}_x$ (the v_1, v_2 and V are *algebraic values* that may be positive or negative). By conservation of momentum,

$$m_1 \vec{v}_1 + m_2 \vec{v}_2 = (m_1 + m_2)\vec{V}$$

or, by factoring out and eliminating $\hat{u}_x$,

$$m_1 v_1 + m_2 v_2 = (m_1 + m_2)V \quad \Rightarrow \quad V = \frac{m_1 v_1 + m_2 v_2}{m_1 + m_2} \tag{1}$$

The change of kinetic energy due to the collision is

$$\Delta E_k = E_{k,after} - E_{k,before} = \frac{1}{2}(m_1 + m_2)V^2 - \left(\frac{1}{2}m_1 v_1^2 + \frac{1}{2}m_2 v_2^2\right).$$

Substituting V from (1), we finally have:

$$\Delta E_k = -\frac{1}{2}\frac{m_1 m_2}{m_1 + m_2}(v_1 - v_2)^2 = -\frac{1}{2}\mu v_{12}^2 \tag{2}$$

where $v_{12} = v_1 - v_2$ is the relative velocity of the masses just before the collision, while μ is the *reduced mass* of the system (see Problem 36). Given that $v_1 \neq v_2$ (otherwise there would be no collision!), it follows from (2) that $\Delta E_k \neq 0$. That is, the kinetic energy of the system is *not* conserved. For example, if $m_1 = m_2$ and $v_1 = -v_2$, Eq. (1) yields $V = 0$, so that the composite mass produced by the collision has no kinetic energy: all the initial kinetic energy of the system has been lost. According to the work-energy theorem, this loss of kinetic energy is mainly due to the work of the internal forces that act during the collision of the two bodies. These forces are responsible for the deformation of the bodies in the course of the collision.

38. *Two carts with equal masses $m = 0.25$ kg are placed on a frictionless horizontal pathway, at the end of which there is a spring of constant $k = 50$ N/m (see figure). The left cart is given an initial velocity $v_0 = 3$ m/s to the right, while the right cart is initially at rest. The two carts collide elastically and the right cart falls on the spring. Find the maximum compression suffered by the spring.*

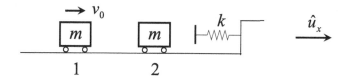

Fig. Problem 38

Solution: The indices 1 and 2 correspond to the left and the right cart, respectively. Let $\vec{v}_1 = v_1 \hat{u}_x$ and $\vec{v}_2 = v_2 \hat{u}_x$ be the velocities of the carts immediately after their elastic collision (the v_1 and v_2 are algebraic values). By conservation of momentum,

$$m v_0 \hat{u}_x + 0 = m v_1 \hat{u}_x + m v_2 \hat{u}_x \quad \Rightarrow \quad v_0 = v_1 + v_2 \tag{1}$$

Kinetic energy is also conserved:

$$\frac{1}{2} m v_0^2 + 0 = \frac{1}{2} m v_1^2 + \frac{1}{2} m v_2^2 \quad \Rightarrow \quad v_0^2 = v_1^2 + v_2^2 \tag{2}$$

By squaring (1) and by taking (2) into account, we find that $v_1 v_2 = 0$. The case $v_2 = 0$ is impossible since (1) would then yield $v_1 = v_0$ (this would mean that, after the collision, cart 1 would go on moving to the right with its initial velocity while cart 2 would remain at rest!). The only possible solution of (1) and (2) is, therefore,

$$v_1 = 0, \quad v_2 = v_0 .$$

That is, cart 2 takes on the velocity of cart 1, which, in turn, comes to a halt.

Let us concentrate now on cart 2 after its collision with cart 1. As is easy to see, the total force on this cart is equal to the force exerted by the compressed spring when the cart and the spring are in contact. Since this force is conservative, the total mechanical energy of cart 2 is constant. Just before colliding with the spring, cart 2 has the velocity v_0 acquired previously, while the spring is uncompressed. After its collision with the spring, cart 2 will stop momentarily at a position where the spring will suffer a maximum deformation, say, Δl. By conservation of mechanical energy,

$$\frac{1}{2} m v_0^2 + 0 = 0 + \frac{1}{2} k (\Delta l)^2 \quad \Rightarrow \quad \Delta l = v_0 \sqrt{\frac{m}{k}} \simeq 0.212 \, m .$$

39. A **ballistic pendulum** *is a device used to measure the speed of a fast-moving projectile, such as a bullet fired from a gun. This device is described in the following example (see figure): A cowboy fires horizontally and from a close distance at a small wooden plate of mass M, hanging from a string at the entrance of a Western saloon. The mass of the bullet is m. The bullet is wedged into the wood, which begins to oscillate like a pendulum, reaching a maximum height h above its initial position. Find the initial speed v of the bullet, as well as the loss of kinetic energy during the collision of the bullet with the plate.*

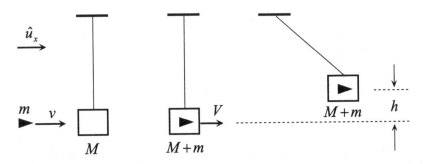

Fig. Problem 39

Solution: We consider the system "bullet + wood". The total momentum of the system just before the collision is the same as the momentum immediately after:

$$m v \hat{u}_x + 0 = (M + m) V \hat{u}_x \quad \Rightarrow \quad m v = (M + m) V \quad \Rightarrow$$

$$v = \left(1 + \frac{M}{m} \right) V \tag{1}$$

where V is the velocity of the composite mass $(M + m)$ right after the collision. The external forces acting on the system after the collision are its weight, $(M + m)g$, which is a conservative force, and the tension of the string, which produces no work since it is always normal to the velocity of the plate (explain this). Thus, the total mechanical energy of the system is constant after the collision. In particular, the mechanical energy right after the collision (when the system has velocity V) equals the mechanical energy at the maximum height h (where the system is momentarily at rest):

$$\frac{1}{2}(M + m)V^2 + 0 = 0 + (M + m)gh \quad \Rightarrow \quad V = \sqrt{2gh} \tag{2}$$

From (1) and (2) we have:

$$v = \left(1 + \frac{M}{m} \right)\sqrt{2gh} \tag{3}$$

By using (2) and (3), we find the loss of kinetic energy during the collision:

$$\Delta E_k = E_{k,after} - E_{k,before} = \frac{1}{2}(M+m)V^2 - \frac{1}{2}mv^2 = -M\left(1+\frac{M}{m}\right)gh.$$

What accounts for this loss of energy?

40. *In figure 1, a bullet of mass m, moving horizontally, falls on a piece of wood of mass M and becomes embedded in it. The wood is connected to a spring of constant k. The coefficient of friction (static and kinetic) between the wood and the horizontal surface is μ, while the spring has initially (i.e., before the collision) its natural length. Find the values of the initial velocity v_0 of the bullet, for which the wood will finally come to rest.*

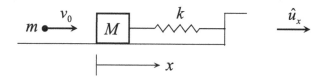

Fig. Problem 40 (1)

Solution: Since the collision is plastic, only the total momentum is conserved:

$$\vec{P}_{justbefore} = \vec{P}_{rightafter} \quad \Rightarrow \quad mv_0\hat{u}_x + 0 = (M+m)V\hat{u}_x \quad \Rightarrow$$

$$mv_0 = (M+m)V \tag{1}$$

where V is the velocity of the system "wood + bullet" right after the collision.

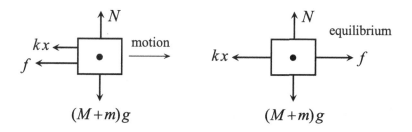

Fig. Problem 40 (2)

The forces on the system $(M+m)$, shown in figure 2, are its weight, $(M+m)g$, the normal reaction N by the horizontal surface, the friction f, and the force kx from the spring (where x is the compression of the spring). The normal forces $(M+m)g$ and N balance each other. When, after the collision, the mass $(M+m)$ moves to the right, the friction f is *kinetic*, directed to the left and equal to $f = \mu N = \mu(M + m)g$. If $(M + m)$ finally balances at rest, the friction f becomes *static*; it is now directed to the right (why?) and satisfies the condition $f \leq \mu N$.

The force kx by the spring is conservative, with potential energy

$$E_p = \frac{1}{2}kx^2.$$

Right after the collision, the system has an initial mechanical energy

$$E_{init} = \frac{1}{2}(M + m)V^2 + 0 \quad \left(\text{since } x = 0, \text{ so that } E_p = 0\right).$$

When the wood comes to rest, having traveled a distance x to the right (so that the compression of the spring at that location is also x), the final mechanical energy of the system is

$$E_{fin} = 0 + \frac{1}{2}kx^2.$$

The change of the mechanical energy of the system within this time interval is equal to the work of the kinetic friction f (the normal reaction N does not produce work). Taking into account that the direction of f is opposite to that of the displacement, we have:

$$\Delta E = E_{fin} - E_{init} = W_f \quad \Rightarrow$$

$$\frac{1}{2}kx^2 - \frac{1}{2}(M + m)V^2 = -fx = -\mu(M + m)gx \quad \Rightarrow$$

$$kx^2 + 2\mu(M + m)gx - (M + m)V^2 = 0.$$

Setting $\lambda = \frac{M+m}{k}$, we write:

$$x^2 + 2\mu\lambda gx - \lambda V^2 = 0 \tag{2}$$

Keeping only the positive root of (2) (since $x > 0$), we have:

$$x = -\mu\lambda g + \sqrt{\mu^2\lambda^2 g^2 + \lambda V^2} \tag{3}$$

Expression (3) gives the distance at which the wood comes to a halt (even if this occurs only momentarily) relative to its initial equilibrium position. Now, in order that the wood remain in equilibrium in this new position, so that it doesn't begin to move backward (to the left), the *static* friction f must be sufficient to balance the push kx by the spring. That is,

$$kx = f \leq f\text{max} = \mu N = \mu(M + m)g \quad \Rightarrow \quad kx \leq \mu\lambda kg \quad \Rightarrow$$

$$x \leq \mu\lambda g \tag{4}$$

Substituting (3) into (4), and solving for V, we find:

$$V \leq \mu g \sqrt{3\lambda} \tag{5}$$

Solving now (1) for V, and substituting the result into (5), we finally have:

$$v_0 \leq \left(1 + \frac{M}{m}\right)\mu g \sqrt{3\lambda} \quad \text{where} \quad \lambda = \frac{M + m}{k}.$$

41. *A body that is initially at rest explodes into two fragments of masses m_1 and m_2.*
    (a) Show that the kinetic energies of the fragments are inversely proportional
    to the masses of the fragments. (b) If Q is the total energy liberated by the
    explosion, find the kinetic energies of the two fragments.

Solution:

(a) This process is in effect equivalent to a time-reversed plastic collision. Since
    the explosion takes place almost instantaneously, and since during this process
    the external forces that act on the system are negligible compared to the internal
    ones, we can consider that the total momentum of the system just before and
    right after the explosion is the same (cf. Sect. 6.6):

$$\vec{P}_{before} = \vec{P}_{after} \quad \Rightarrow \quad 0 = \vec{p}_1 + \vec{p}_2 \quad \Rightarrow \quad \vec{p}_1 = -\vec{p}_2$$

where $\vec{p}_1$ and $\vec{p}_2$ are the momenta of the two fragments. If p_1 and p_2 are the magnitudes
of the momenta, then $p_1 = p_2 \equiv p$. Now, in general, the kinetic energy of a body of
mass m and momentum p is equal to

$$E_k = \frac{p^2}{2m}.$$

Hence,

$$E_{k,1} = \frac{p_1^2}{2m_1} = \frac{p^2}{2m_1}, \quad E_{k,2} = \frac{p_2^2}{2m_2} = \frac{p^2}{2m_2}$$

and, by dividing these relations,

$$\frac{E_{k,1}}{E_{k,2}} = \frac{m_2}{m_1} \tag{1}$$

(b) Assume now that

$$E_{k,1} + E_{k,2} = Q.$$

By applying a well-known property of proportions to (1), we have:

$$\frac{E_{k,1}}{E_{k,1} + E_{k,2}} = \frac{m_2}{m_1 + m_2}, \quad \frac{E_{k,1} + E_{k,2}}{E_{k,2}} = \frac{m_1 + m_2}{m_1}.$$

Thus,

$$E_{k,1} = \frac{m_2}{m_1 + m_2}Q, \quad E_{k,2} = \frac{m_1}{m_1 + m_2}Q.$$

Notice that, if $m_1 << m_2$, then $E_{k,2} \approx 0$ and $E_{k,1} \approx Q$. That is, almost all of the liberated energy Q goes off as kinetic energy of the smaller fragment. This is what happens, for example, in a radioactive decay of an atomic nucleus.

Problems for Chap. 7

42. *A wheel of mass M and radius R may rotate about a horizontal pivot passing through its center, O (see figure). A string wound around the wheel is connected to an object of mass m. As the string unwinds, the mass m moves downward while the wheel rotates. Find (a) the linear acceleration, a, of m; (b) the angular acceleration, α, of the wheel; (c) the tension F of the string; and (d) the reaction N by the pivot at O. (The moment of inertia of the wheel, relative to the pivot, is $I = \frac{1}{2}MR^2$.)*

Solution: The mass m performs purely translational motion, while the motion of the wheel M is purely rotational. For the translational motion we choose the positive direction downward. For the rotational motion we choose the positive direction counterclockwise. Equivalently, if z is the axis of rotation, directed normal to the page, its positive direction is *outward* (i.e., toward the reader), in accordance with the right-hand rule.

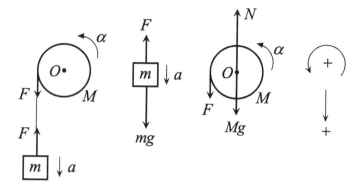

Fig. Problem 42

The equation of translational motion of m is

$$mg - F = ma \qquad (1)$$

Since the wheel M does not perform translational motion,

$$Mg + F - N = 0 \quad \Rightarrow \quad N = Mg + F \qquad (2)$$

The equation of rotational motion of M is $\Sigma T_z = I\alpha$, where α is the angular acceleration of the wheel and ΣT_z is the z-component of the total torque relative to O (equal to the sum of the z-components of the torques of all forces acting on M). The Mg and N pass through O, thus they produce no torque relative to that point. The total torque on M, therefore, is due to the tension F of the string and is directed positively, since F tends to turn the wheel counterclockwise. We have [cf. Sect. (7.5)]:

$$\sum T_z = \sum (R F_T) = FR$$

so that

$$FR = I\alpha \quad \Rightarrow \quad F = \frac{I}{R}\alpha \qquad (3)$$

We now notice that the velocity of the downward motion of m is equal, in magnitude, to the linear velocity of revolution, $R\omega$, of all points of the circumference of the wheel (where ω is the instantaneous angular velocity of the wheel). Similarly, the linear acceleration, a, of m equals the *tangential* acceleration, $R\alpha$, of all points of the circumference of the wheel. That is,

$$a = R\alpha \qquad (4)$$

From (3) and (4) we have:

$$F = \frac{I}{R^2}a = \frac{1}{R^2}(\frac{1}{2}MR^2)a \quad \Rightarrow \quad F = \frac{1}{2}Ma \tag{5}$$

Substituting (5) into (1), we find:

$$a = \frac{2mg}{2m + M} \tag{6}$$

Equation (4) then yields:

$$\alpha = \frac{a}{R} = \frac{2mg}{(2m + M)R}.$$

By (5) and (6) we get:

$$F = \frac{mMg}{2m + M} \tag{7}$$

Finally, (2) and (7) yield:

$$N = Mg\left(1 + \frac{m}{2m + M}\right).$$

What will happen if the mass of the wheel is negligible $(M = 0)$?

43. *The string in the figure is wound around the wheel. We are given the mass M and the radius R of the wheel, as well as the masses m_1, m_2 of the two objects hanging from the ends of the string, where $m_1 < m_2$. Find the linear accelerations of the two masses, the angular acceleration, α, of the wheel, and the tensions F_1, F_2 of the string. (The moment of inertia of the wheel is $I = \frac{1}{2}MR^2$.)*

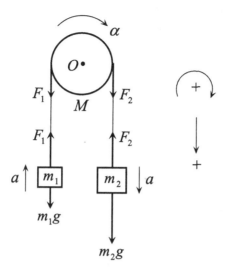

Fig. Problem 43

Solution: For the translational motion of m_1 and m_2 we take the positive direction downward. For the rotational motion of M we choose the positive direction clockwise. That is, we consider that the z-axis of rotation, which is normal to the page, is positively directed *into* the page. We arbitrarily assume that the wheel is accelerating clockwise and, correspondingly, the masses m_1, m_2 accelerate as shown in the figure. If our assumption is correct, we will find positive values for the magnitudes of the accelerations at the end; otherwise these values will be found negative.

We note that, in contrast to Problem 11 where the tension of the string was constant along the string and was not affected by the pulley, here the string is *wound* around the wheel without gliding on it, which allows the two sections of the string to develop independent tensions F_1 and F_2. The equations of motion for m_1 and m_2 are

$$m_1g - F_1 = -m_1a \quad \Rightarrow \quad F_1 - m_1g = m_1a \tag{1}$$

$$m_2g - F_2 = m_2a \tag{2}$$

Notice that the linear accelerations of m_1 and m_2 are equal in magnitude and equal to the *tangential* acceleration, $R\alpha$, of all points of the circumference of the wheel. That is,

$$a_1 = a_2 \equiv a = R\alpha \tag{3}$$

(the a and α represent *magnitudes*; thus they are positive). The equation of rotational motion of M is $\Sigma T_z = I\alpha$. Of the forces acting on the wheel, only the tensions F_1 and F_2 produce torque relative to O (what are the remaining forces?). The torque of

F_2 is positive, while that of F_1 is negative (why?). We have:

$$\sum T_z = \sum (RF_T) = F_2R - F_1R = (F_2 - F_1)R.$$

Thus, by using (3),

$$(F_2 - F_1)R = I\alpha \Rightarrow F_2 - F_1 = \frac{I}{R}\alpha = \frac{I}{R^2}a = \frac{1}{R^2}(\frac{1}{2}MR^2)a \Rightarrow$$

$$F_2 - F_1 = \frac{1}{2}Ma \tag{4}$$

Adding (1) and (2), solving for $(F_2 - F_1)$, and comparing with (4), we find:

$$a = \frac{(m_2 - m_1)g}{m_1 + m_2 + \frac{M}{2}} \quad \text{and} \quad \alpha = \frac{a}{R} = \frac{(m_2 - m_1)g}{\left(m_1 + m_2 + \frac{M}{2}\right)R}.$$

We notice that $a > 0$ and $\alpha > 0$ (since $m_1 < m_2$). Thus the assumed direction for the accelerations was correct. From (1) and (2) we easily find the tensions F_1 and F_2 (this is left as an exercise). If the wheel is considered almost *massless* ($M = 0$), then, as we can show,

$$a = \frac{(m_2 - m_1)g}{m_1 + m_2}, \quad F_1 = F_2 = \frac{2m_1m_2}{m_1 + m_2}g.$$

The device described in this problem is called *Atwood's machine* and is useful in the study of uniformly accelerated motion, as well as in the experimental measurement of the acceleration of gravity, g. By choosing the masses m_1 and m_2 so that their difference $(m_2 - m_1)$ is very small, we achieve a value a of the acceleration that is small enough to be easily measurable.

44. *A wooden disk of mass M and radius R may rotate without friction about a horizontal pivot passing through the center O of the disk (see figure). A bullet of mass m and horizontal velocity v hits the disk at its top point A and is wedged there, setting the disk in rotation. Find the velocity and the acceleration of the bullet at point B, where the angle AOB is $\pi/2$. (The moment of inertia of the disk is $I = \frac{1}{2}MR^2$.)*

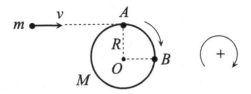

Fig. Problem 44

Solution: We choose as positive the clockwise direction of rotation. That is, we consider that the z-axis of rotation (pivot), which is normal to the page, is positively directed *into* the page (this is, by definition, the direction of the unit vector $\hat{u}_z$). We call ω_A and ω_B the angular velocities of the disk when the bullet is at A (right after the collision) and B, respectively. Also, we call v_A and v_B the linear velocities of the bullet at these points. Obviously,

$$v_A = R\omega_A, \quad v_B = R\omega_B \tag{1}$$

Consider the system "bullet + disk". Its angular momentum relative to O, just before the collision, equals its angular momentum right after (cf. Sect. 6.6):

$$\vec{L}_{before} = \vec{L}_{after} \quad \Rightarrow \quad mRv\hat{u}_z + 0 = mRv_A\hat{u}_z + I\omega_A\hat{u}_z \quad \Rightarrow$$

$$mRv = mRv_A + I\omega_A = mR^2\omega_A + \frac{1}{2}MR^2\omega_A \quad \Rightarrow$$

$$\omega_A = \frac{2mv}{(2m + M)R} \tag{2}$$

From (1) and (2) $\Rightarrow$

$$v_A = \frac{2mv}{2m + M} \tag{3}$$

(Note that, after the bullet is embedded in the disk, the center of mass of the system does *not* coincide with the fixed point O but rotates with the disk.) To find ω_B we will use conservation of mechanical energy (remember that there is no friction at O). For the gravitational potential energy, we take $E_p = 0$ at the horizontal plane passing through O. The external forces on the system are the weights mg and Mg and the reaction N from the pivot. The force Mg acts at the center of mass O of the disk and corresponds to a potential energy $E_{p,M} = 0$. The force mg corresponds to a potential energy $E_{p,m} = mgy$, where y is the instantaneous vertical distance of m from the horizontal plane passing through O. The force N does not produce work ($W_N = 0$) since its point of application, O, remains fixed. The change of total mechanical energy of the system is given by $\Delta E = W_N = 0 \Rightarrow E = $ constant. We may thus apply conservation of mechanical energy at the two instants when the bullet is at A and at B, taking into account relations (1):

$$E_A = \frac{1}{2}mv_A^2 + \frac{1}{2}I\omega_A^2 + mgR = \frac{1}{2}\left(m + \frac{M}{2}\right)R^2\omega_A^2 + mgR,$$

$$E_B = \frac{1}{2}mv_B^2 + \frac{1}{2}I\omega_B^2 + 0 = \frac{1}{2}\left(m + \frac{M}{2}\right)R^2\omega_B^2.$$

Equating $E_A = E_B$, we find:

$$\omega_B = \left[\omega_A^2 + \frac{4mg}{(2m + M)R} \right]^{1/2} \tag{4}$$

where ω_A is given by (2). By the second of relations (1) we can then evaluate v_B.

We now want to find the acceleration a_B of m at B. The bullet m performs circular motion of radius R. If α_B is the angular acceleration of the disk when m is at B, we have [see Eqs. (2.36) and (2.37)]:

$$a_B = |\vec{a}_B| = \sqrt{a_T^2 + a_N^2} \quad \text{where} \quad a_T = R\alpha_B, \quad a_N = R\omega_B^2.$$

Thus,

$$a_B = R\sqrt{\alpha_B^2 + \omega_B^4} \tag{5}$$

To find α_B, we use the relation

$$\sum T_z = I_{tot}\alpha_B \tag{6}$$

where ΣT_z is the z-component of the total external torque relative to O when m is at B (equal to the sum of z-components of the torques of all forces acting on the system) and where I_{tot} is the *total* moment of inertia of the system "bullet + disk". The forces Mg and N pass through O, thus they produce no torque. Therefore, the total torque on the system is due to mg alone and is directed positively, since mg tends to rotate the disk clockwise:

$$\sum T_z = mgR \quad \text{when } m \text{ is at } B.$$

To find I_{tot}, we think as follows: We imagine that the disk M is partitioned into elementary masses m_i at distances r_i from the axis of rotation. Given that the bullet m is at a distance R from the axis, we have:

$$I_{tot} = \sum_i m_i r_i^2 + mR^2 = I + mR^2 = \frac{1}{2}MR^2 + mR^2 = \left(m + \frac{M}{2} \right)R^2.$$

Equation (6) now yields:

$$mgR = \left(m + \frac{M}{2} \right)R^2\alpha_B \quad \Rightarrow \quad \alpha_B = \frac{2mg}{(2m + M)R} \tag{7}$$

By substituting (4) and (7) into (5), we find a_B.

Exercise: Suppose now that the bullet is not wedged into the disk but just bounces at A, giving the disk an initial angular velocity ω_A. What will then be the angular velocity ω_B and the angular acceleration α_B of the disk when the point of impact moves to location B? (Notice that, now, $I_{tot} = I$. Is there an external torque?)

45. *A wooden rod of length l and mass M may rotate without friction about a horizontal pivot passing through one end of the rod (see figure). The rod is initially vertical and at rest. A bullet of mass m, moving horizontally, falls on the center C of the rod and becomes embedded in the wood. (a) Find the velocity v of the bullet for which the rod will reach the horizontal position (where it will stop momentarily). (b) Find the angular and the linear acceleration of the center C of the rod when the rod is at the horizontal position. (The moment of inertia of the rod is $I = \frac{1}{3}Ml^2$.)*

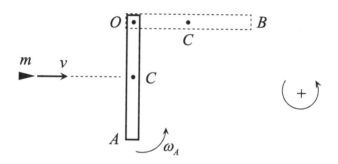

Fig. Problem 45

Solution: We define the positive direction of rotation to be counterclockwise. That is, the z-axis of rotation (pivot), which is normal to the page, is positively directed *outward* (this is also the direction of $\hat{u}_z$). We call ω_A the angular velocity of the rod right after the collision, when the rod is still vertical (position A). At the final, horizontal position B the rod stops momentarily, which means that $\omega_B = 0$. We work as in Problem 44. We thus consider the system "bullet + rod". By applying conservation of angular momentum during the collision, relative to the fixed end O of the rod, we have:

$$\vec{L}_{before} = \vec{L}_{after} \;\Rightarrow\; mv\left(\frac{l}{2}\right)\hat{u}_z + 0 = mv_A\left(\frac{l}{2}\right)\hat{u}_z + I\omega_A\hat{u}_z$$

where v_A is the velocity of the bullet right after the collision. Substituting for I, and using the fact that $v_A = \frac{1}{2}\omega_A$, we get:

$$\frac{1}{2}mlv = \left(\frac{m}{4} + \frac{M}{3}\right)l^2\omega_A \;\Rightarrow\; v = \frac{3m + 4M}{6m}l\omega_A \qquad (1)$$

To find ω_A we apply conservation of mechanical energy. We choose the zero level of gravitational potential energy at height $l/2$ below O. The external forces on the system are the weights mg and Mg and the reaction N from the pivot. The potential energy of the system is

$$E_p = E_{p,m} + E_{p,M} = mgy + Mgy = (m + M)gy$$

where y is the instantaneous vertical distance of C from the horizontal plane of zero potential energy [cf. Sect. 7.9, Eq. (7.66)]. The reaction N produces no work, thus the mechanical energy of the system is constant. At the extreme positions A and B, we have (taking into account that $y_A = 0$, $y_B = l/2$ and $\omega_B = 0$):

$$E_A = \frac{1}{2} I_{tot} \omega_A^2 + (m + M)gy_A = \frac{1}{2}\left(\frac{m}{4} + \frac{M}{3}\right)l^2 \omega_A^2 + 0,$$

$$E_B = \frac{1}{2} I_{tot} \omega_B^2 + (m + M)gy_B = 0 + (m + M)g\frac{l}{2},$$

where we have put

$$I_{tot} = \frac{1}{3}Ml^2 + m\left(\frac{l}{2}\right)^2 = \left(\frac{m}{4} + \frac{M}{3}\right)l^2.$$

By equating $E_A = E_B$, we find:

$$\omega_A = \left[\frac{12(m + M)g}{(3m + 4M)l}\right]^{1/2} \tag{2}$$

Substituting (2) into (1), we have:

$$v = \frac{1}{6m}\left[12(m + M)(3m + 4M)gl\right]^{1/2}.$$

The angular acceleration α_B at B is found by using the relation

$$\sum T_z = I_{tot}\alpha_B = \left(\frac{m}{4} + \frac{M}{3}\right)l^2\alpha_B \tag{3}$$

The reaction N produces no torque with respect to O, in contrast to the weights mg and Mg that cause torque in the negative direction (since they tend to rotate the rod clockwise). The mg and Mg act at the center C of the rod. At the horizontal position B of the rod,

$$\sum T_z = -mg\frac{l}{2} - Mg\frac{l}{2} = -\frac{1}{2}(m + M)gl.$$

Substituting this into (3), we get the algebraic value of the angular acceleration at position B:

$$\alpha_B = -\frac{6(m+M)g}{(3m+4M)l} \tag{4}$$

Now, the point C performs circular motion of radius $l/2$. The magnitude of its linear acceleration at B is

$$a_B = \frac{l}{2}\left(\alpha_B^2 + \omega_B^4\right)^{1/2} = \frac{l}{2}|\alpha_B| \quad \text{(since } \omega_B = 0\text{)}.$$

Substituting for α_B from (4), we finally have:

$$a_B = \frac{3(m+M)g}{3m+4M}.$$

46. *A homogeneous sphere of mass M and radius R is let to roll from the top of an inclined plane of angle θ and maximum height h. Find the velocity of the sphere at the moment when it reaches the base of the incline, by using two methods: (a) conservation of mechanical energy; (b) the equations of motion of the sphere. Also, find the minimum value of the coefficient of static friction, μ, in order that the sphere may roll without slipping on the incline. (The moment of inertia of the sphere is $I = \frac{2}{5}MR^2$.)*

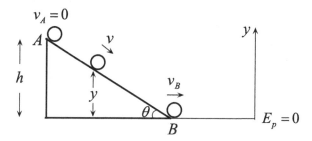

Fig. Problem 46 (1)

Solution: Consider an arbitrary position of the sphere at height y above the base of the incline (see figure 1). Call ω the angular velocity of rotation of the sphere and let v be the velocity of the center of mass of the sphere at that position. Since the sphere executes pure rolling (without slipping),

$$v = R\omega \tag{1}$$

The total mechanical energy of the sphere at the considered position is

$$E = \frac{1}{2}Mv^2 + \frac{1}{2}I\omega^2 + Mgy.$$

The value of E is constant during the motion. Indeed, the forces on the sphere are its weight Mg (to which there corresponds the potential energy $E_p = Mgy$), the normal reaction N from the incline, and the *static* friction f that prevents slipping. The reaction N produces no work, being normal to the velocity v. The friction f also doesn't produce work, given that the motion is pure rolling (see Sect. 7.11). Hence,

$$\Delta E = W_N + W_f = 0 \quad \Rightarrow \quad E = \text{constant}.$$

In particular, for the extreme positions A and B we have that $E_A = E_B$. Explicitly,

$$0 + 0 + Mgh = \frac{1}{2}Mv_B^2 + \frac{1}{2}I\omega_B^2 + 0.$$

Substituting for I, and using the fact that, by (1), $v_B = R\omega_B$, we have:

$$gh = \frac{1}{2}v_B^2 + \frac{1}{5}v_B^2 = \frac{7}{10}v_B^2 \quad \Rightarrow \quad v_B = \sqrt{\frac{10}{7}gh} \qquad (2)$$

The result is independent of M, R, θ. It is interesting to note that, if in place of the sphere we had, e.g., a cube *sliding without friction* on the inclined plane, the corresponding velocity at the base of the incline would be $v_B' = \sqrt{2gh} > v_B$. This is understood by noting that, in the case of the sphere, part of the total kinetic energy is used as kinetic energy of rotation ($I\omega^2/2$) at the expense of translational kinetic energy ($Mv^2/2$), and thus at the expense of the final velocity v_B of the sphere.

Alternatively, we can use the equations of motion of the sphere. The forces acting on it are the weight Mg, the normal reaction N from the incline, and the static friction f (see figure 2). The friction f is directed toward the top A of the incline in order to prevent slipping.

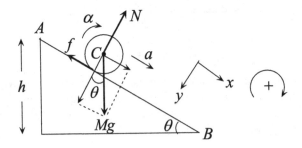

Fig. Problem 46 (2)

Consider an arbitrary position of the sphere on the incline. Call a the linear acceleration of the center of mass C of the sphere and let α be the angular acceleration of rotation of the sphere. The z-axis of rotation, which is moving down the incline with the sphere, passes through C and is normal to the page and directed *inward* (that is, the positive direction of rotation is clockwise). The equations of translational motion of the sphere are

$$\sum F_x = M a_x, \quad \sum F_y = M a_y, \quad \text{where} \quad a_x = a, \quad a_y = 0.$$

Explicitly,

$$Mg\sin\theta - f = Ma \tag{3}$$

$$Mg\cos\theta - N = 0 \quad \Rightarrow \quad N = Mg\cos\theta \tag{4}$$

The equation of rotation about the center of mass C is $\Sigma T_z = I\alpha$. The only force producing torque about C is the static friction f (given that Mg and N pass through C). This torque is positively directed since f tends to rotate the sphere clockwise. We thus have:

$$fR = \left(\frac{2}{5}MR^2\right)\alpha \Rightarrow f = \frac{2}{5}MR\alpha = \frac{2}{5}Ma \tag{5}$$

where we have used the pure-rolling condition $a = R\alpha$. Substituting (5) into (3) and solving for a, we find:

$$a = \frac{5}{7}g\sin\theta \tag{6}$$

(Note that a is independent of R.) Substituting, now, (6) into (5), we have:

$$f = \frac{2}{7}Mg\sin\theta \tag{7}$$

Taking into account that f is *static* friction (since there is no slipping), and using (4) and (7), we have:

$$f \leq f_{\max} = \mu N \quad \Rightarrow \quad \frac{2}{7}Mg\sin\theta \leq \mu Mg\cos\theta \quad \Rightarrow$$

$$\mu \geq \frac{2}{7}\tan\theta \quad \Leftrightarrow \quad \mu_{\min} = \frac{2}{7}\tan\theta \tag{8}$$

Note that $\mu_{\min} = 0$ when $\theta = 0$. This means that *the sphere may perform pure rolling on a <u>horizontal</u> surface without the aid of static friction*. In other words, *pure rolling is possible on a frictionless horizontal surface* (see also Problem 47).

We also need to find the velocity v_B of the sphere at the base B of the inclined plane. Along the incline (i.e., along the x-axis) the center C of the sphere executes uniformly accelerated motion with constant acceleration a given by (6). We thus use the formula

$$v^2 = v_0^2 + 2a(x - x_0)$$

with $v = v_B$, $v_0 = v_A = 0$ and $x - x_0 = AB = \frac{h}{\sin\theta}$:

$$v_B^2 = 0 + 2\left(\frac{5}{7}g\sin\theta\right)\frac{h}{\sin\theta} = \frac{10}{7}gh \quad \Rightarrow \quad v_B = \sqrt{\frac{10}{7}gh},$$

in accordance with (2).

If in place of the sphere we had a *homogeneous cylinder* of radius R ($I = MR^2/2$), relations (2), (6) and (8) would read, respectively,

$$v_B = \sqrt{\frac{4}{3}gh}, \quad a = \frac{2}{3}g\sin\theta, \quad \mu \geq \frac{1}{3}\tan\theta$$

(show this!). We notice that the acceleration a of the sphere along the incline is greater than that of the cylinder. We also notice that the accelerations of both these objects are independent of their respective radii R. Hence, if a sphere and a cylinder start to roll simultaneously from the top of an inclined plane, the sphere will arrive at the base of the incline first, *regardless of the dimensions of the two objects!*

47. *A disk of radius R is at rest on a frictionless ($\mu = 0$) horizontal surface. (a) At what vertical distance h above the center of mass O of the disk must we apply a constant horizontal force F in order for the disk to roll without slipping? (b) Show that free rolling (F = 0) on a horizontal surface is possible without the aid of static friction. (The moment of inertia of the disk is $I = \frac{1}{2}MR^2$.)*

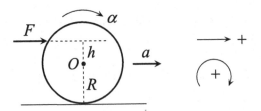

Fig. Problem 47

Solution: The positive directions for translational and rotational motion are indicated in the figure. The z-axis of rotation, which is moving with the disk, passes through O and is normal to the page and directed *into* it. We call M the mass of the disk and we denote by a and α the linear and the angular acceleration, respectively, of the disk. The forces on the disk are its weight Mg, the normal reaction N by the horizontal surface (equal in magnitude to the weight), and the horizontal force F that we apply (there is no friction). The force F is the total force on the disk. The total torque relative to O is due to F alone, given that Mg and N pass through O. The equations of translational and rotational motion are:

$$\sum \vec{F} = M\vec{a} \Rightarrow F = Ma \tag{1}$$

$$\sum T_z = I\alpha \quad \Rightarrow \quad Fh = \left(\frac{1}{2}MR^2\right)\alpha = \frac{1}{2}MRa \tag{2}$$

where we have used the pure-rolling condition $a = R\alpha$. Dividing (2) by (1), we have:

$$h = \frac{R}{2} \quad \text{for pure rolling without friction.}$$

Note that, for $F \neq 0$, the result is independent of F and M. In the case of *free rolling* ($F = 0$) relations (1) and (2) yield $a = 0$ and $\alpha = 0$. That is, *the disk rolls with constant velocity and without slipping on the frictionless horizontal surface*.

48. *A disk of radius R and mass M rolls without slipping on a horizontal surface under the action of a constant horizontal force F (see figure).*

1. *Find the values of the vertical distance h above the center of mass O of the disk, at which we must apply F in order for the static friction f exerted on the disk to be directed (a) to the left, (b) to the right.*
2. *For the values $h = 0$ and $h = R$, find the values of the static coefficient of friction, μ, in order that the disk may perform pure rolling.*

(The moment of inertia of the disk is $I = \frac{1}{2}MR^2$.)

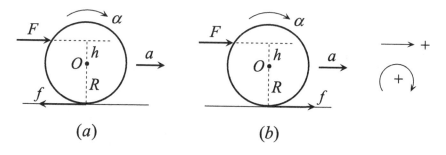

Fig. Problem 48

Solution: The positive directions for translational and rotational motion are indicated in the figure (the moving z-axis of rotation is normal to the page and directed *into* it). We call a the linear acceleration and α the angular acceleration of the disk. The pure-rolling condition is

$$a = R\alpha \tag{1}$$

The total force on the disk is $F \pm f$ (the vertical forces, i.e., the weight Mg and the normal reaction N by the surface, balance each other). The forces F and f are the only ones that produce torque relative to O, given that the vertical forces pass through O. The equations of translational and rotational motion are:

$$\sum \vec{F} = M\vec{a}, \quad \sum T_z = I\alpha \tag{2}$$

We apply relations (2) separately for cases (a) and (b), taking into account condition (1) and the fact that the torque of F relative to O is positive, while the torque of f is positive in case (a) and negative in case (b).

(a) The static friction f is directed to the left:

$$F - f = Ma \tag{3}$$

$$Fh + fR = \left(\frac{1}{2}MR^2\right)\alpha = \frac{1}{2}MRa \tag{4}$$

Dividing (4) by (3), we find:

$$\frac{3R}{2}f = \left(\frac{R}{2} - h\right)F \tag{5}$$

Given that $f > 0$ and $F > 0$ (the f and F represent *magnitudes*), we must have:

$$\frac{R}{2} - h > 0 \quad \Rightarrow \quad 0 \leq h < \frac{R}{2}.$$

(b) The static friction f is directed to the right:

$$F + f = Ma \tag{6}$$

$$Fh - fR = \left(\frac{1}{2}MR^2\right)\alpha = \frac{1}{2}MRa \tag{7}$$

Dividing (7) by (6), we find:

$$\frac{3R}{2}f = \left(h - \frac{R}{2}\right)F \tag{8}$$

This time we must have

$$h - \frac{R}{2} > 0 \quad \Rightarrow \quad \frac{R}{2} < h \le R.$$

We now seek the static coefficient of friction μ for pure rolling when $h = 0$ and $h = R$. Setting $h = 0$ in (5) and $h = R$ in (8), we find that, in both cases,

$$f = \frac{F}{3}.$$

We thus have:

$$f \le f_{\text{max}} = \mu N = \mu Mg \quad \Rightarrow \quad \frac{F}{3} \le \mu Mg \quad \Rightarrow \quad \mu \ge \frac{F}{3Mg}.$$

Exercise: Show that, when $F = 0$ (free rolling), then $f = 0$, $a = 0$ and $\alpha = 0$. That is, the disk rolls freely with constant velocity, without the aid of static friction (see also Problem 47).

49. *A disk of radius R and mass M has a string wound around it. A mass m is attached to the end of the string, as shown in figure 1. As the string is unwound, the mass m moves downward while the disk rolls without slipping on the horizontal surface. Find the accelerations a_1 and a_2 of the disk and the mass, the tension F of the string, and the static friction f (magnitude <u>and</u> direction) at the point of contact O. (The moment of inertia of the disk is $I = \frac{1}{2} MR^2$.)*

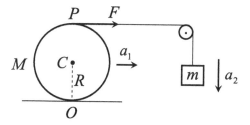

Fig. Problem 49 (1)

Solution: Since we don't yet know the direction of the friction f, we will arbitrarily assume that it is to the right. If we find a negative magnitude for f at the end, our assumption will prove to be incorrect. The acceleration a_1 of the disk is, by definition, the acceleration of its center of mass C relative to the horizontal surface, or, equivalently, relative to the point of contact O: $a_1 = a_{C,O}$. The acceleration a_2 of m with respect to the horizontal surface (directed downward) is equal in magnitude to the acceleration of the highest point P of the disk, relative to the surface (or, relative to O): $a_2 = a_{P,O}$. In Sect. 7.10 we showed that $a_{P,O} = 2a_{C,O}$. Therefore,

$$a_2 = 2a_1 \tag{1}$$

We set

$$a_1 = a, \quad a_2 = 2a.$$

If α is the angular acceleration of the disk relative to C, the condition for pure rolling demands that

$$a_1 = a = R\alpha \tag{2}$$

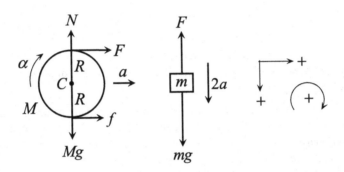

Fig. Problem 49 (2)

In figure 2 we draw the forces on each body separately. The positive directions for translation and rotation are as indicated in the figure (the moving z-axis of rotation of the disk is normal to the page and directed *into* it). The total force on M is $(F + f)$ (since $N = Mg$), while that on m is $(mg\text{-}F)$. The Mg and N do not produce torque relative to C, since they both pass through C. Thus, the total torque on M relative to C is

$$\sum T_z = FR - fR = (F - f)R$$

(notice that the torque of F is positive, while that of f is negative). We write the equations of translational motion for m and for M, as well as the equation of rotational motion for M:

$$mg - F = m(2a) = 2ma \tag{3}$$

$$F + f = Ma \tag{4}$$

$$(F - f)R = I\alpha = \left(\frac{1}{2}MR^2\right)\alpha \overset{(2)}{\Rightarrow} F - f = \frac{1}{2}Ma \qquad (5)$$

Dividing (4) by (5) $\Rightarrow$ $F = 3f$ (6)

From (4) and (6) $\Rightarrow$ $f = \frac{Ma}{4}$ (7)

From (6) and (7) $\Rightarrow$ $F = \frac{3Ma}{4}$ (8)

From (3) and (8) $\Rightarrow$ $a = a_1 = \frac{4mg}{8m+3M}$ (9) $\checkmark$

From (1) and (9) $\Rightarrow$ $a_2 = \frac{8mg}{8m+3M}$ (10) $\checkmark$

From (7–9) $\Rightarrow$ $f = \frac{Mmg}{8m+3M}$, $F = \frac{3Mmg}{8m+3M}$ (11) $\checkmark$

We notice that $f > 0$; thus, the arbitrarily chosen direction for f was correct. Equations (9–11) constitute the solution to the problem.

50. *A student is sitting on a stool that can rotate about a frictionless vertical axis. The student is holding, with both arms extended, a pair of dumbbells, each of mass m, while the stool is rotating with an initial angular velocity ω_1. The student now pulls the dumbbells closer to his body, so that their initial distance R_1 from the axis of rotation decreases to R_2. (a) Find the new angular velocity ω_2 of the stool. Assume, approximately, that the moment of inertia of the system "stool + student" (<u>without</u> the dumbbells) remains unchanged, equal to I_0. (b) Compare the initial with the final kinetic energy of the system (<u>with</u> the dumbbells).*

Solution: Consider the system "stool + student + dumbbells". Its initial and its final moment of inertia is, respectively,

$$I_1 = I_0 + mR_1^2 + mR_1^2 = I_0 + 2mR_1^2,$$

$$I_2 = I_0 + 2mR_2^2.$$

The angular momentum of the system, relative to any point of the z-axis of rotation (which is a principal axis), is directed parallel to the axis. Its initial and its final value is, respectively,

$$L_1 = I_1\omega_1 = \left(I_0 + 2mR_1^2\right)\omega_1,$$
$$L_2 = I_2\omega_2 = \left(I_0 + 2mR_2^2\right)\omega_2.$$

The external forces on the system are the weights of the bodies and the reaction of the floor on the stool. Since all of these forces are in the vertical direction, none of them produces torque about the axis of rotation. Hence, $\Sigma T_z = 0$, so that the angular momentum of the system is constant:

$$L_1 = L_2 \quad \Rightarrow \quad I_1\omega_1 = I_2\omega_2 \quad \Rightarrow$$

$$\omega_2 = \frac{I_1}{I_2}\omega_1 = \frac{I_0 + 2mR_1^2}{I_0 + 2mR_2^2}\omega_1 .$$

Note that $\omega_2 > \omega_1$, given that $R_1 > R_2$.

The initial and the final kinetic energy of the system is, respectively, $E_{k,1} = \frac{1}{2}I_1\omega_1^2$ and $E_{k,2} = \frac{1}{2}I_2\omega_2^2$. Thus,

$$\frac{E_{k,2}}{E_{k,1}} = \frac{I_2}{I_1}\left(\frac{\omega_2}{\omega_1}\right)^2 = \frac{I_2}{I_1}\left(\frac{I_1}{I_2}\right)^2 = \frac{I_1}{I_2} > 1 .$$

That is, $E_{k,2} > E_{k,1}$. Given that the external forces produce no work (why?) the increase of total kinetic energy is due to the work of the *internal* forces (specifically, the work done by the student in pulling the dumbbells closer to his body).

Problems for Chap. 8

51. *A homogeneous sphere of density ρ_1 is let to sink inside a vessel containing a liquid of density ρ (where $\rho < \rho_1$), as seen in the figure. The height of the free surface of the liquid above the bottom of the vessel is h. (a) Find the time t_1 it will take the sphere to reach the bottom. (b) If a second sphere of density $\rho_2 > \rho_1$ is let to sink simultaneously with the first sphere, which sphere will reach the bottom first?*

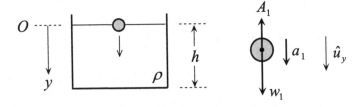

Fig. Problem 51

Solution:

(a) Call m_1, V_1, w_1 the mass, the volume and the weight, respectively, of the first sphere. Then, $m_1 = \rho_1 V_1$ and

$$w_1 = m_1 g = \rho_1 g V_1 .$$

The buoyant force on the sphere is equal to the weight of the displaced liquid:

$$A_1 = \rho g V_1 .$$

The total force on the sphere is

$$\vec{F}_1 = \vec{w}_1 + \vec{A}_1 = w_1 \hat{u}_y + (-A_1 \hat{u}_y) = (w_1 - A_1)\hat{u}_y = (\rho_1 - \rho)g V_1 \hat{u}_y .$$

By Newton's law,

$$\vec{F}_1 = m_1 \vec{a}_1 = m_1 a_1 \hat{u}_y = (\rho_1 V_1 a_1)\hat{u}_y .$$

Hence,

$$\rho_1 V_1 a_1 = (\rho_1 - \rho)g V_1 \quad \Rightarrow \quad a_1 = \left(1 - \frac{\rho}{\rho_1}\right)g \qquad (1)$$

Since $a_1 =$ constant, the motion of the sphere is uniformly accelerated. Therefore,

$$y = y_0 + v_0 t + \frac{1}{2}a_1 t^2$$

where, in our case, $y_0 = 0$, $v_0 = 0$, $y = h$ and $t = t_1$. Thus,

$h = \frac{1}{2}a_1 t_1^2 \quad \Rightarrow \quad t_1 = \sqrt{\frac{2h}{a_1}}$, where a_1 is given by (1).

(b) Similarly, for the second sphere we have:

$$t_2 = \sqrt{\frac{2h}{a_2}} \quad \text{where} \quad a_2 = \left(1 - \frac{\rho}{\rho_2}\right)g .$$

Given that $\rho_1 < \rho_2$, we see that $a_1 < a_2$, so that $t_1 > t_2$. That is, *the denser sphere ρ_2 will reach the bottom first, regardless of the dimensions of the two spheres!*

52. *Consider a metal sphere of radius $R = 5$ cm and density $\rho_\sigma = 3.1$ g/cm³ (see figure). We want to cover it with a wooden spherical shell of density $\rho_\xi = 0.7$ g/cm³, so that the composite spherical structure will float on water ($\rho = 1$ g/cm³) or at least will not sink in it. Find the minimum thickness x of the wooden shell.*

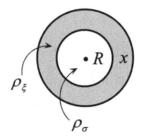

Fig. Problem 52

Solution: Call m_σ and V_σ the mass and the volume of the metal sphere, and m_ξ and V_ξ the mass and the volume of the wooden shell. The mass and the volume of the composite sphere are $m_{tot} = m_\sigma + m_\xi$ and $V_{tot} = V_\sigma + V_\xi$. In order for that sphere to not sink, its *average* density ρ_σ' must not exceed the density of water: $\rho_\sigma' \leq \rho$, where

$$\rho_\sigma' = \frac{m_{tot}}{V_{tot}} = \frac{m_\sigma + m_\xi}{V_{tot}} = \frac{\rho_\sigma V_\sigma + \rho_\xi V_\xi}{V_{tot}} = \frac{\rho_\sigma V_\sigma + \rho_\xi (V_{tot} - V_\sigma)}{V_{tot}} \Rightarrow$$

$$\rho_\sigma' = \rho_\xi + (\rho_\sigma - \rho_\xi)\frac{V_\sigma}{V_{tot}} .$$

Therefore,

$$\rho_\sigma' \leq \rho \quad \Rightarrow \quad (\rho_\sigma - \rho_\xi)\frac{V_\sigma}{V_{tot}} \leq \rho - \rho_\xi \tag{1}$$

But,

$$V_\sigma = \frac{4}{3}\pi R^3 , \quad V_{tot} = \frac{4}{3}\pi (R + x)^3 \quad \text{and} \quad \frac{V_\sigma}{V_{tot}} = \left(\frac{R}{R + x}\right)^3 .$$

From (1) we then have:

$$\left(\frac{R}{R + x}\right)^3 \leq \frac{\rho - \rho_\xi}{\rho_\sigma - \rho_\xi} \Rightarrow \frac{R + x}{R} \geq \left(\frac{\rho_\sigma - \rho_\xi}{\rho - \rho_\xi}\right)^{1/3} \Rightarrow$$

$$x \geq \left[\left(\frac{\rho_\sigma - \rho_\xi}{\rho - \rho_\xi}\right)^{1/3} - 1\right] R .$$

After making numerical substitutions, we find: $x_{min} = 5$ cm.

53. *In Sect. 5.4 we learned that, if we suspend a mass m from a spring of constant k, the mass will be able to execute vertical oscillations about its equilibrium position with period $T = 2\pi \sqrt{m/k}$. We now place the system into a liquid of*

density ρ smaller than the average density of the suspended mass (this means that m would sink if it were not held by the spring). Show that the period of oscillation remains unchanged inside the liquid.

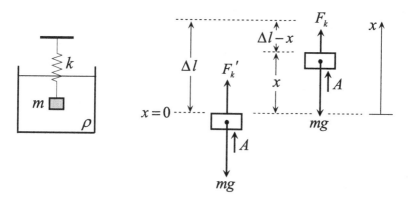

Fig. Problem 53

Solution: The symbols in the figure are the same as in Sect. 5.4(*b*). The mass *m* is initially in equilibrium at $x = 0$, where the spring is extended by Δl relative to its natural length and exerts a force $F_k' = k\Delta l$. The mass is then displaced a distance x above the equilibrium position and is now subject to a force $F_k = k(\Delta l - x)$ by the spring. The only new element is the constant buoyant force $A = \rho g V$, where V is the volume of the mass *m*. Notice that $A < mg$, since the density ρ of the liquid is smaller than that of the immersed mass. At the equilibrium position $x = 0$, we have:

$$F_k' + A - mg = 0 \quad \Rightarrow \quad k\Delta l + A - mg = 0 \tag{1}$$

The *resultant* force on *m* at position *x* is

$$F = F_k + A - mg = k(\Delta l - x) + A - mg \overset{(1)}{\Rightarrow} F = -kx \tag{2}$$

According to Eq. (2), the mass *m* may execute harmonic oscillations about the equilibrium position *x*=0, with period $T = 2\pi\sqrt{m/k}$. We note that this period is independent of the density ρ of the medium that surrounds the oscillating mass.

54. *A cylinder of height h and average density ρ_k floats with its axis vertical, partially immersed in a liquid of density ρ (where $\rho > \rho_k$). The cylinder is subjected to a small vertical displacement from its equilibrium position and is then let free. Show that the cylinder will execute harmonic oscillations about its equilibrium position and find the angular frequency ω. As an application, find ω for the case where the cylinder is half-immersed in the liquid at the state of equilibrium.*

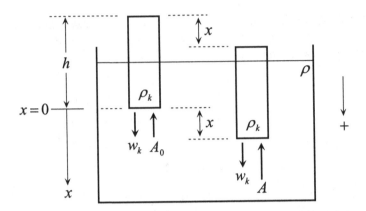

Fig. Problem 54

Solution: In the figure we see the cylinder at the equilibrium position ($x = 0$) and at an arbitrary position a vertical distance x below. Let V_k be the volume of the cylinder. Its weight is then $w_k = \rho_k \, g V_k$. We call V_0 and V the volumes of the displaced liquid at the considered positions $x = 0$ and $x > 0$, respectively. The corresponding buoyant forces on the cylinder at these positions are

$$A_0 = \rho g V_0, \quad A = \rho g V.$$

If S is the area of the base of the cylinder, then $V_k = S \, h$. Also,

$$V = V_0 + Sx \quad \Leftrightarrow \quad V - V_0 = Sx.$$

At the equilibrium position ($x = 0$) we have $w_k = A_0 \Rightarrow$

$$\rho_k V_k = \rho V_0 \tag{1}$$

The *resultant* force on the cylinder at the arbitrary position x is

$$F = w_k - A = \rho_k g V_k - \rho g V = (\rho_k V_k - \rho V)g$$

or, by substituting for $\rho_k V_k$ from (1),

$$F = (\rho V_0 - \rho V)g = -\rho(V - V_0)g = -\rho(Sx)g \Rightarrow$$
$$F = -(\rho g S)x \equiv -kx \tag{2}$$

According to (2), the cylinder may execute vertical harmonic oscillations about the equilibrium position ($x = 0$) with angular frequency $\omega = \sqrt{k/m}$, where m is the mass of the cylinder. Setting $m = \rho_k V_k = \rho_k Sh$ and $k = \rho g S$, we find:

$$\omega = \sqrt{\frac{\rho g}{\rho_k h}} \tag{3}$$

In the special case where $V_0 = V_k /2$, relation (1) yields $\rho / \rho_k = V_k / V_0 = 2$, so that from (3) we have: $\omega = \sqrt{2g/h}$.

Exercise: Find the period T and the frequency f of oscillation. Would the oscillation be perfectly harmonic if we had a sphere in place of the cylinder? A cube?

55. *A goldsmith tries to sell a crown to a king, assuring him that it is golden and has no internal cavities. Before he makes a decision, the king seeks the advice of Archimedes, who weighs the crown immediately and finds its weight to be $w = 7.84$ N. Archimedes then weighs the crown while it is immersed in water, measuring an <u>apparent</u> weight $w_\varphi = 6.86$ N. Knowing that the density of water is $\rho = 10^3$ kg/m³, while that of gold is $\rho_x = 19.3 \times 10^3$ kg/m³, what does Archimedes conclude regarding the authenticity of the crown?*

Solution: In essence, Archimedes is seeking the average density ρ' of the crown in order to compare it with the density ρ_x of gold. If V is the volume of the crown, the *actual* weight of the crown is

$$w = \rho' g V \tag{1}$$

We have an equation with two unknowns, namely, ρ' and V. Assuming that we do not have the experimental means needed in order to determine the volume V directly, we seek a second equation to eliminate that volume from the problem. The *apparent* weight of the crown, when the latter is immersed in the water, is equal to its true weight *minus* the buoyant force A:

$$w_\varphi = w - A = \rho' g V - \rho g V \Rightarrow$$

$$w_\varphi = (\rho' - \rho) g V \tag{2}$$

Dividing (2) by (1) and solving for ρ', we find:

$$\rho' = \frac{w}{w - w_\varphi} \rho = 8 \times 10^3 \, kg/m^3 < \rho_x \,.$$

Thus, either the crown has internal cavities or it is not made of pure gold (The historians of the period give no information regarding the fate of the goldsmith...).

56. *A tank containing water is placed on a scale, which shows a total weight w. What will be the new reading of the scale if we add (a) a stone of volume V, totally immersed in the water and kept at rest suspended by a string from a fixed support? (b) a stone of volume V that is sinking freely in the water? (c) a piece of wood of weight w', floating on the surface of the water? (d) a wooden cube*

of weight w', performing vertical oscillations of angular frequency ω about its equilibrium position, at which position it floats at rest? (The density ρ of water is assumed known.)

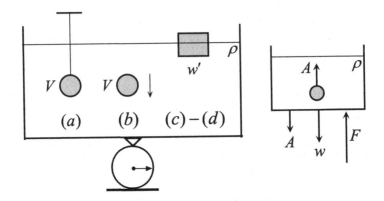

Fig. Problem 56

Solution: In all cases the water exerts a buoyant force A on the immersed body, directed upward (see figure). By the action-reaction law the body exerts, in turn, a force on the surrounding water, of equal magnitude A and directed *downward*. The external forces on the system "tank + water" are its weight w, the reaction A to the buoyant force by the immersed body, and the force F on the tank by the scale. The force F is the one that corresponds to the reading of the scale. Since the system "tank + water" is in equilibrium (note that the immersed body is *not* a part of the considered system!), the following condition must be valid in all cases:

$$F = w + A \tag{1}$$

For cases (a) and (b) we may argue as follows: Whether suspended and at rest or sinking freely, the stone is subject to the same buoyant force, $A = \rho g V$, from the water. Thus, in both these cases the stone exerts the same downward reaction A to the surrounding liquid. For cases (a) and (b), then, relation (1) yields:

$$F = w + \rho g V .$$

In case (c), since the floating piece of wood is in equilibrium, its weight w' is equal in magnitude to the buoyant force: $A = w'$. Equation (1) then gives:

$$F = w + w' .$$

Regarding case (d), we note the following: When the wooden cube floats in equilibrium, the buoyant force on it is $A_0 = w'$. When the cube oscillates vertically (see also Problem 54) the buoyant force changes periodically with time, according to an equation of the form:

$$A = A_0 + B \cos \omega t = w' + B \cos \omega t$$

where B is a constant quantity. Equation (1) then yields:

$$F = w + w' + B \cos \omega t \,.$$

57. *A container of uniform cross-sectional area A is filled with a liquid of density ρ. On the wall of the container, at a vertical distance h below the free surface of the liquid, we open a hole of area a, where a ≪ A. Find the initial velocity of outflow of the liquid from the hole, as well as the volume flow rate of the initial outflow (The liquid is considered ideal.).*

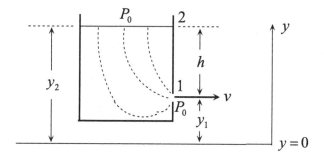

Fig. Problem 57

Solution: As soon as we open the hole, a tube of flow (a few streamlines of which are shown in the figure) is formed, extending from the free surface of the liquid (cross-section $A_2 = A$) to the hole (cross-section $A_1 = a$). The hydrostatic pressure at both these cross-sections is equal to the atmospheric pressure: $P_1 = P_2 = P_0$. We call $v_1 = v$ and v_2 the flow velocities at the considered cross-sections, and we call y_1, y_2 the heights at which these cross-sections are located above an arbitrary horizontal reference level. By the equation of continuity,

$$A_1 v_1 = A_2 v_2 \quad \Rightarrow \quad v_2 = \frac{A_1}{A_2} v_1 = \frac{a}{A} v \quad \Rightarrow \quad v_2 \simeq 0 \qquad (1)$$

since $a \ll A$, by assumption, so that $a/A \simeq 0$. By Bernoulli's equation,

$$P_1 + \frac{1}{2}\rho v_1^2 + \rho g y_1 = P_2 + \frac{1}{2}\rho v_2^2 + \rho g y_2 \overset{(1)}{\Rightarrow}$$

$$P_0 + \frac{1}{2}\rho v^2 + \rho g y_1 = P_0 + \rho g y_2 \Rightarrow v^2 = 2g(y_2 - y_1) \quad \Rightarrow$$

$$v = \sqrt{2gh}.$$

Notice that the velocity of initial outflow is independent of the density ρ of the liquid. The volume flow rate is

$$\Pi = av = a\sqrt{2gh}.$$

Exercise: Do the problem again, this time assuming that the area a of the hole is *not* negligible compared to the cross-sectional area A of the container. Show that the exact value of the initial velocity of outflow is

$$v = \left[\frac{2gh}{1 - \left(\frac{a}{A}\right)^2} \right]^{1/2} \quad \left(\simeq \sqrt{2gh} \text{ when } \frac{a}{A} << 1 \right).$$

58. *A horizontal tube of cross-sectional area A_1 narrows to a coaxial tube of cross-section A_2 ($A_2 < A_1$), as shown in the figure. A vertical tube is attached at position 1 of the wider section, while another vertical tube is attached at position 2 of the narrower section. Water flows inside this horizontal coaxial system of tubes, while inside the vertical tubes the (static) water rises to corresponding heights h_1 and h_2. (a) In which of the two vertical tubes does the water rise higher? (b) Find the volume flow rate of the horizontal flow, given the A_1, A_2, h_1, h_2.*

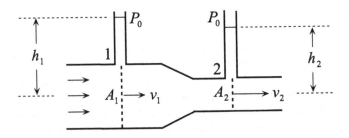

Fig. Problem 58

Solution: We assume that the water columns in the two vertical tubes are measured, approximately, from the axis of the horizontal system of tubes. Since the water is static inside the vertical tubes, we can use the fundamental equation of Hydrostatics to find the pressures at the cross-sections 1 and 2:

$$P_1 = P_0 + \rho g h_1, \ P_2 = P_0 + \rho g h_2 \tag{1}$$

where ρ is the density of water and P_0 is the atmospheric pressure. Now, as we showed in Sect. 8.13, in horizontal flow the pressure is greater where the cross-section of the tube of flow is greater. Given that $A_1 > A_2$, we must have $P_1 > P_2$. It thus follows from relations (1) that $h_1 > h_2$. That is, the water rises to a greater height at the vertical tube attached to the *wider* section of the coaxial system.

By the equation of continuity,

$$A_1 v_1 = A_2 v_2 \quad \Rightarrow \quad v_2 = \frac{A_1}{A_2} v_1 \tag{2}$$

where v_1, v_2 are the flow velocities at the two cross-sections. By Bernoulli's equation for horizontal flow,

$$P_1 + \frac{1}{2}\rho v_1^2 = P_2 + \frac{1}{2}\rho v_2^2 \tag{3}$$

Substituting (1) and (2) into (3), and solving for v_1, we find:

$$v_1 = A_2 \left[\frac{2g(h_1 - h_2)}{A_1^2 - A_2^2} \right]^{1/2}.$$

The flow rate is

$$\Pi = A_1 v_1 = A_1 A_2 \left[\frac{2g(h_1 - h_2)}{A_1^2 - A_2^2} \right]^{1/2}.$$

59. *At the bottom of a very large tank we open a hole of area $a = 1$ cm^2 (which is considered negligible compared to the cross-sectional area of the tank). We now place the tank under a faucet from which water flows at a volume rate of $\Pi = 140$ cm^3/s. At what height h will the free surface of the water rise in the tank? ($g = 980$ cm/s^2).*

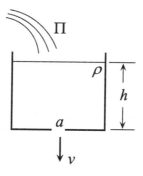

Fig. Problem 59

Solution: There is an incoming flow of water in the tank, of constant volume flow rate Π (see figure). At the same time there is an outgoing flow through the hole. The level of the water in the tank is initially rising, since the outgoing flow rate, equal to $a.v$ (where v is the velocity of outflow from the hole) is less than the incoming flow rate Π. The water level will stop rising as soon as the two flow rates become equal:

$$\Pi = av \tag{1}$$

Now, the hole is at a vertical distance h below the free surface of water in the tank. As we showed in Problem 57, the outflow velocity for a given (albeit now variable) h is

$$v = \sqrt{2gh} \tag{2}$$

(it makes no difference that this time the hole is at the bottom, rather than on the wall, of the tank). From (1) and (2) we have:

$$h = \frac{\Pi^2}{2ga^2} = 10\,cm.$$

60. *Water is running from a faucet of cross-sectional area A, at a constant volume flow rate Π. Find the cross-sectional area of the column of water at a vertical distance h below the faucet, as a function of Π, A and h.*

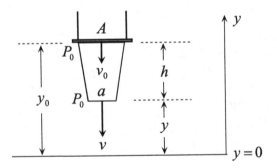

Fig. Problem 60

Solution: Let a be the cross-sectional area of the column of water at distance h below the faucet; let v be the flow velocity at that cross-section; and let y be the height at which the cross-section a is located above an arbitrary horizontal reference level (see figure). At the "mouth" of the faucet (cross-section A) the corresponding quantities are v_0 and y_0. The pressure at both cross-sections A and a is equal to the atmospheric pressure P_0. We apply the laws of Hydrodynamics at these cross-sections, assuming that all conditions of ideal flow are satisfied.

By the equation of continuity,

$$\Pi = A v_0 = a v \tag{1}$$

By Bernoulli's equation,

$$P_0 + \frac{1}{2}\rho v_0^2 + \rho g y_0 = P_0 + \frac{1}{2}\rho v^2 + \rho g y \quad \Rightarrow$$

$$v^2 = v_0^2 + 2g(y_0 - y) \quad \Rightarrow \quad v = \sqrt{v_0^2 + 2gh} \tag{2}$$

From (1) $\Rightarrow$

$$v_0 = \frac{\Pi}{A} \tag{3}$$

and

$$a = \frac{\Pi}{v} \tag{4}$$

where the flow rate Π is assumed known. By substituting (2) and (3) into (4), we finally have:

$$a = \frac{\Pi}{\sqrt{\left(\frac{\Pi}{A}\right)^2 + 2gh}} \,.$$

Mathematical Supplement

1. Differential of a Function

Consider a function $y = f(x)$. Let Δx be an arbitrary change of the independent variable, from its initial value x to $x + \Delta x$. The corresponding change of y is

$$\Delta y = f(x + \Delta x) - f(x).$$

Note that Δy is a function of *two* independent variables, x and Δx.
The *derivative* of f at a point x is defined as

$$f'(x) = \lim_{\Delta x \to 0} \frac{f(x + \Delta x) - f(x)}{\Delta x} = \lim_{\Delta x \to 0} \frac{\Delta y}{\Delta x} \tag{1}$$

It follows from (1) that a function $\varepsilon(x, \Delta x)$ must exist such that

$$\frac{\Delta y}{\Delta x} = f'(x) + \varepsilon(x, \Delta x) \quad \text{where} \quad \lim_{\Delta x \to 0} \varepsilon(x, \Delta x) = 0 \tag{2}$$

Thus,

$$\Delta y = f'(x)\Delta x + \varepsilon(x, \Delta x)\Delta x \tag{3}$$

The product $f'(x)\Delta x$ is *linear* (i.e., of the first degree) in Δx, while the product $\varepsilon(x, \Delta x)\Delta x$ must only contain terms that are *at least of the second degree* in Δx (that is, it may not contain a constant or a linear term). We write, symbolically,

$$\varepsilon(x, \Delta x)\Delta x \equiv O(\Delta x^2) \quad \text{where} \quad \Delta x^2 \equiv (\Delta x)^2 \, (\neq \Delta(x^2)!).$$

Equation (3) is then written as

© The Editor(s) (if applicable) and The Author(s), under exclusive license
to Springer Nature Switzerland AG 2020
C. J. Papachristou, *Introduction to Mechanics of Particles and Systems*,
https://doi.org/10.1007/978-3-030-54271-9

$$\boxed{\Delta y = f'(x)\Delta x + O(\Delta x^2)} \tag{4}$$

We notice that Δy is the sum of a linear and a higher-order term in Δx. Furthermore, the derivative of f at x is the coefficient of Δx in the linear term.

Example Let $y = f(x) = x^3$. Then,

$$\Delta y = f(x + \Delta x) - f(x) = (x + \Delta x)^3 - x^3 = 3x^2 \Delta x + \left(3x\Delta x^2 + \Delta x^3\right),$$

by which we have that $f'(x) = 3x^2$ and $O(\Delta x^2) = 3x\Delta x^2 + \Delta x^3$.

The linear term in (4), which is a function of x and Δx, is called the *differential* of the function $y = f(x)$ and is denoted dy:

$$\boxed{dy = df(x) = f'(x)\Delta x} \tag{5}$$

Equation (4) is then written:

$$\Delta y = dy + O(\Delta x^2) \tag{6}$$

If Δx is *infinitesimal* ($|\Delta x| \ll 1$) we can make the approximation $O(\Delta x^2) \simeq 0$. Hence,

$$\Delta y \simeq dy = f'(x)\Delta x \quad \text{for infinitesimal } \Delta x \tag{7}$$

Careful, however: for *finite* (i.e., non-infinitesimal) Δx, the *difference* Δy and the *differential dy* are, in general, separate quantities!

An exception to the above general remark occurs in the case of *linear* functions. Let $y = f(x) = ax + b$. Then,

$$\Delta y = f(x + \Delta x) - f(x) = [a(x + \Delta x) + b] - (ax + b) = a\Delta x$$

and

$$dy = f'(x)\Delta x = (ax + b)'\Delta x = a\Delta x = \Delta y .$$

That is, *for linear functions (and only for such functions) the differential dy is the same as the difference* Δy, even if these quantities assume finite values. This means that, for these functions, $O(\Delta x^2) = 0$.

Let us see a few applications of the definition (5) of the differential:

$$\text{For } f(x) = x^a \quad \Rightarrow \quad d(x^a) = (x^a)'\Delta x = ax^{a-1}\Delta x ;$$
$$\text{for } f(x) = e^x \quad \Rightarrow \quad d(e^x) = (e^x)'\Delta x = e^x \Delta x ;$$

$$\text{for } f(x) = \ln x \quad \Rightarrow \quad d(\ln x) = (\ln x)'\Delta x = \frac{1}{x}\Delta x.$$

In particular, for $f(x) = x$, we have: $dx = (x)'\Delta x = 1 \cdot \Delta x \Rightarrow$

$$\boxed{\Delta x = dx} \tag{8}$$

in accordance with the remark made earlier regarding linear functions. Equation (5) may thus be rewritten in a more symmetric form:

$$\boxed{dy = df(x) = f'(x)dx} \tag{9}$$

By dividing the above relation by dx, we can express the derivative as follows:

$$\boxed{f'(x) = \frac{dy}{dx} = \frac{df(x)}{dx}} \tag{10}$$

In words: *The derivative of a function is equal to the differential of the function divided by the differential (or, the change) of the independent variable.*

2. Differential Operators

We introduce a notation that proves to be important in higher mathematics:

$$\frac{df(x)}{dx} \equiv \frac{d}{dx}f(x) \tag{11}$$

Note that this notation tries to "mimic" the properties of ordinary multiplication of numbers:

$$\frac{\alpha \cdot \beta}{\gamma} = \frac{\alpha}{\gamma} \cdot \beta$$

except that the expression $\frac{d}{dx}$ is *not* a number! The symbol $\frac{d}{dx}$ is called a *differential operator* and, when placed in front of a function $f(x)$, it *instructs* us to take the derivative of $f(x)$. We thus write:

$$\boxed{f'(x) = \frac{df(x)}{dx} = \frac{d}{dx}f(x)} \tag{12}$$

Relation (12) contains three different notations for the derivative of a function.

Higher-order derivatives can also be expressed in terms of differential operators. Thus, the second derivative of $y = f(x)$ is written:

$$f''(x) = \frac{d}{dx}\frac{df(x)}{dx} = \frac{d}{dx}\left(\frac{d}{dx}f(x)\right) = \left(\frac{d}{dx}\right)^2 f(x) = \frac{d^2}{dx^2}f(x)$$

or

$$f''(x) = \frac{d^2 f(x)}{dx^2} = \frac{d^2 y}{dx^2} \tag{13}$$

Exercise: Verify the following properties of the differential:

1. $d[f(x) \pm g(x)] = df(x) \pm dg(x)$
2. $d[f(x)g(x)] = f(x)dg(x) + g(x)df(x)$
3. $d[cf(x)] = cdf(x)$ $c=const.$
4. $d\left[\frac{f(x)}{g(x)}\right] = \frac{g(x)df(x) - f(x)dg(x)}{[g(x)]^2}$

3. Geometrical Significance of the Differential

We make the *special assumption* that the quantities x and y are *dimensionless* and, moreover, equal lengths on the x- and y-axes correspond to *equal changes* of x and y.

Figure MS.1 shows a section of the graph of a function $y = f(x)$. We consider an arbitrary point $M \equiv (x, y)$ of the curve and we draw the tangent line to this curve at that point. This line forms an angle θ with the x-axis. As we see in the figure, to the change $\Delta x = MA$ of x there corresponds the change $\Delta y = AM'$ of y. The linear section AB, then, represents the differential dy of f for the given values of x and Δx. Indeed, taking into account that $f'(x) = \tan\theta$, we have:

Fig. MS.1 Graph of a function $y = f(x)$ and the tangent line to it at a point M

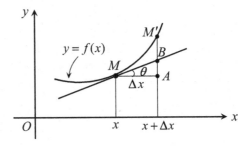

$$dy = f'(x)\Delta x = (\tan\theta)\Delta x = \frac{AB}{MA}MA = AB.$$

Also, by Eq. (6),

$$O(\Delta x^2) = \Delta y - dy = AM' - AB = BM'.$$

If the function f is *linear*, then $B \equiv M'$ so that $O(\Delta x^2) = 0$ and $\Delta y = dy$.

4. Derivative of a Composite Function

Consider two functions f and g such that $y = f(u)$ and $u = g(x)$. The *composite function* $(f \circ g)$ is defined as follows:

$$y = (f \circ g)(x) \equiv f[g(x)].$$

To simplify our notation, we write $y = y(u)$, $u = u(x)$ and $y = y(x) = y[u(x)]$.
We want to find an expression for the derivative of y with respect to x. This derivative is equal to the quotient dy/dx. We write:

$$y'(x) = \frac{dy}{dx} = \frac{dy}{du}\frac{du}{dx} = y'(u)u'(x),$$

which expresses the familiar "chain rule" for calculating the derivative of a composite function.

5. Differential Equations

In the case of an *algebraic equation* the solution is a set of *numbers* (real or complex). On the other hand, the solution of a *differential equation* is a set of *functions*. A first-order differential equation is an algebraic relation involving the derivative of an unknown function as well as, possibly, the function itself. An equation of this kind admits an *infinite* number of solutions. In order to specify a particular solution, in addition to the differential equation one must also be provided with an *initial condition*. Let us see a few examples:

1. Determine the function $y = y(x)$ that satisfies the differential equation

$$y' = e^{2x} \tag{14}$$

as well as the initial condition

$$y = 1 \quad \text{for} \quad x = 0 \tag{15}$$

Solution: Equation (14) is written:

$$\frac{dy}{dx} = e^{2x} \Rightarrow dy = e^{2x} dx \tag{16}$$

We can proceed in two ways:

(a) Take the *indefinite* integral of (16) in order to find the *general solution* of (14):

$$\int dy = \int e^{2x} dx \Rightarrow y + C_1 = \frac{1}{2} e^{2x} + C_2 \Rightarrow y = \frac{1}{2} e^{2x} + (C_2 - C_1)$$

or, since the constants C_1, C_2 are arbitrary,

$$y = \frac{1}{2} e^{2x} + C \tag{17}$$

Note that the general solution (17) represents an infinite set of functions corresponding to the various values of the arbitrary constant C. By applying the initial condition (15) to (17), we can determine the value of the constant C :

$$1 = \frac{1}{2} e^0 + C \quad \Rightarrow \quad C = \frac{1}{2}.$$

By (17) we then find the *particular solution*

$$y = \frac{1}{2} (e^{2x} + 1).$$

(b) Take the *definite* integrals of the two sides of (16), putting as lower limits the corresponding values of y and x given by the initial condition (15) (as upper limits simply place the variables y and x):

$$\int_1^y dy = \int_0^x e^{2x} dx \quad \Rightarrow \quad y - 1 = \left[\frac{1}{2} e^{2x} \right]_0^x = \frac{1}{2} e^{2x} - \frac{1}{2} e^0 \Rightarrow y = \frac{1}{2} (e^{2x} + 1).$$

Note that this second way is the shortest if we are only interested in finding a particular solution of the differential equation.

2. Solve the differential equation $y' = 5y$, with the initial condition: $y = 3$ for $x = 0$.

Solution:

$$\frac{dy}{dx} = 5y \quad \Rightarrow \quad \frac{dy}{y} = 5dx \Rightarrow \int_3^y \frac{dy}{y} = 5\int_0^x dx \quad \Rightarrow \quad [\ln y]_3^y = 5x \quad \Rightarrow$$

$$\ln\left(\frac{y}{3}\right) = 5x \Rightarrow y = 3e^{5x}.$$

3. Solve the equation $y' = \frac{2}{2x+1}$, with the initial condition: $y = 0$ for $x = 0$.

Solution:

$$\frac{dy}{dx} = \frac{2}{2x+1} \quad \Rightarrow \quad dy = \frac{2dx}{2x+1} \quad \Rightarrow \quad \int_0^y dy = 2\int_0^x \frac{dx}{2x+1} \quad \Rightarrow$$

$$y = 2\left[\frac{1}{2}\ln(2x+1)\right]_0^x = \ln(2x+1).$$

4. Solve the equation $y' = 2\,xy$, with the initial condition: $y = 5$ for $x = 1$.

Solution:

$$\frac{dy}{dx} = 2xy \quad \Rightarrow \quad \frac{dy}{y} = 2xdx \quad \Rightarrow \quad \int_5^y \frac{dy}{y} = 2\int_1^x xdx \quad \Rightarrow$$

$$\ln\left(\frac{y}{5}\right) = x^2 - 1 \quad \Rightarrow \quad y = 5e^{x^2-1}.$$

Exercise: Find the particular solutions of the differential equations of Examples (2–4), this time by first finding the general solutions of the equations and then applying the initial conditions to determine the values of the corresponding constants, as in Example 1(*a*).

Basic Integrals

$$\int dx = x + C$$

$$\int x^a dx = \frac{x^{a+1}}{a+1} + C \quad (a \neq -1)$$

$$\int \frac{dx}{x} = \ln |x| + C$$

$$\int e^x dx = e^x + C$$

$$\int \cos x \, dx = \sin x + C$$

$$\int \sin x \, dx = -\cos x + C$$

$$\int \frac{dx}{\cos^2 x} = \tan x + C$$

$$\int \frac{dx}{\sin^2 x} = -\cot x + C$$

$$\int \frac{dx}{\sqrt{1-x^2}} = \arcsin x + C$$

$$\int \frac{dx}{1+x^2} = \arctan x + C$$

© The Editor(s) (if applicable) and The Author(s), under exclusive license
to Springer Nature Switzerland AG 2020
C. J. Papachristou, *Introduction to Mechanics of Particles and Systems*,
https://doi.org/10.1007/978-3-030-54271-9

$$\int \frac{dx}{x^2 - 1} = \frac{1}{2} \ln\left|\frac{x - 1}{x + 1}\right| + C$$

$$\int \frac{dx}{\sqrt{x^2 \pm 1}} = \ln\left(x + \sqrt{x^2 \pm 1}\right) + C$$

Index

© The Editor(s) (if applicable) and The Author(s), under exclusive license
to Springer Nature Switzerland AG 2020
C. J. Papachristou, *Introduction to Mechanics of Particles and Systems*,
https://doi.org/10.1007/978-3-030-54271-9

Printed in the United States
by Baker & Taylor Publisher Services